ESE 2025

UPSC ENGINEERING SERVICES EXAMINATION

Preliminary Examination

General Studies and Engineering Aptitude

Basics of Project Management

Comprehensive Theory *with* Practice Questions *and* ESE Solved Questions

MADE EASY PUBLICATIONS

www.madeeasypublications.org

MADE EASY Publications Pvt. Ltd.

Corporate Office: 44-A/4, Kalu Sarai (Near Hauz Khas Metro Station), New Delhi-110016
E-mail: infomep@madeeasy.in
Contact: 9021300500

Visit us at: www.madeeasypublications.org

ESE 2025 Preliminary Examination: Basics of Project Management

1st Edition : 2016
2nd Edition : 2017
3rd Edition: 2018
4th Edition: 2019
5th Edition: 2020
6th Edition: 2021
7th Edition: 2022
8th Edition: 2023

9th Edition: 2024

Preface

B. Singh (Ex. IES)

The compilation of this book **Basics of Project Management** was motivated by the desire to provide a concise book which can benefit students to understand the concepts of this specific topic of General Studies and Engineering Aptitude section.

This textbook provides all the requirements of the students, i.e. comprehensive coverage of theory, fundamental concepts and objective type questions articulated in a lucid language. The concise presentation will help the readers grasp the theory of this subject with clarity and apply them with ease to solve objective questions quickly. This book not only covers the syllabus of ESE in a holistic manner but is also useful for many other competitive examinations. All the topics are given the emphasis they deserve so that mere reading of the book clarifies all the concepts.

We have put in our sincere efforts to present detailed theory and MCQs without compromising the accuracy of answers. For the interest of the readers, some notes, do you know and interesting facts are given in the comprehensive manner. At the end of each chapter, sets of practice question are given with their keys and detailed explanations, that will allow the readers to evaluate their understanding of the topics and sharpen their question solving skills.

Our team has made their best efforts to remove all possible errors of any kind. Nonetheless, we would highly appreciate and acknowledge if you find and share with us any printing and conceptual errors.

It is impossible to thank all the individuals who helped us, but we would like to sincerely thank all the authors, editors and reviewers for putting in their efforts to publish this book.

With Best Wishes

B. Singh

CMD, MADE EASY Group

Basics of Project Management

Contents

Chapter 5

PLANNING PROCESS :

Chapter 6

EXECUTION PROCESS:

Chapter 7

PROJECT CLOSURE & TERMINATION:

Chapter 8

1 Basic Concepts and Project Life Cycle

1.1 WHAT IS PROJECT?

A project is accomplished by performing a set of interrelated activities over a fixed period within certain cost and specifications. Projects are temporary and unique.

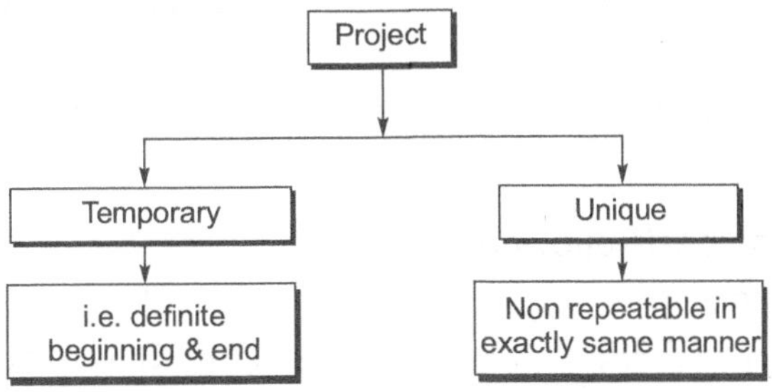

Fig. *Feature of a project*

- Temporariness of projects indicates that they are time bound and have a definite beginning and definite end. It has fixed time span.
- Uniqueness of projects means that they have be created for the first time and are not repeatative or routine.

Project has been defined in various ways. Some of popular definitions are:

- *According to* ***Little and Mirrlees,*** *"A project is any scheme or a part of scheme for investing resources which can be reasonably analysed and evaluated as an independent unit."*
- *According to* ***Harison,*** *"A project can be defined as a non-routine, non-repetitive, one-off undertaking, normally with discrete time, financial and technical performance goal.*

According to Project Management Institutes PMBOK guide.

"A project is a temporary endeavours undertaken to create a unique product, service or result".

Example of projects:

- Development of a software for a business.
- Construction of a building.
- Setting up a power plant.
- Expansion of sales in a new geographic market, etc.

1.2 CLASSIFICATION OF PROJECTS (TAXONOMY OF PROJECTS)

Projects can be classified under different heads as follows:

1. **Classification based on type of work**
   (a) Industrial projects
   (b) Non-industrial projects
      - Infrastructure projects
      - Irrigation projects
      - Building projects
      - Educational projects

- Health care projects
- Advertising and marketing projects
- IT and computer projects
- military and aerospace projects
- Event management projects
- Bank projects

2. Classification on the basis of Project Completion Time
   (a) Long duration projects (Completion time more than 5 years)
   (b) Medium duration projects (3 to 5 years)
   (c) Short duration projects (1 to 3 years)
   (d) Special short duration (Less than 1 year)

3. Classification on the basis of project value
   (a) Mega value projects (> Rs. 1000 crores)
   (b) Large value projects (Rs. 100 crores to 1000 crores)
   (c) Medium value projects (Rs. 1 crores to 100 crores)
   (d) Small value projects (Cost less than 1 crore)

4. Classification on the basis of ownership
   (a) Private sector projects
   (b) Public sector projects
   (c) Joint sector project (PPP)
   (d) Government sector projects

5. Based on the basis of pace of work
   (a) Normal track projects
   (b) Fast track projects

Which of the following are project?

1.	Reading a New Book	Yes
2.	Maintaining of New constructed Road	No
3.	Mass Production in a Car Plant	No
4.	Designing a New Product	Yes

1.2.1 Project Vs Operations

Operations is an ongoing work effort following repetitive process & producing same output or service.

Similarities of Project and operation are:

1. Both consume time and resources
2. Both are executed by individuals
3. Both has some similar phases i.e. planning execution & control
4. Both are executed to meet strategic & organisational objective

Differences of project and operation are:

Project	Operations
Temporary, Unique	Ongoing, repetitive
In Project more skills, are required	It requires limited skills as the process is repetitive
Risk are high at start as no. of unknown and unpredictable elements are present	Risk are not high as it is repetition of activities
A project has definite start and end.	Operations don't have a definite end
Risk are evaluated continually	Here processes are designed to minimise risk.

NOTE : Point of view also plays important role to distinguish project and operation. For customer work effort is project whereas for organisation it is operation as they do it all time.

1.3 WHAT IS MANAGEMENT?

Management is an act of getting people together to accomplish a work or goal using available resources efficiently and effectively.

Management process involves: In the project management context, a project management process consists of a linear sequence of processes:

- Initiation process
- Planning process
- Execution process
- Monitoring and control process
- Closing process

There are various type of resources within management:

(a) Human resources
(b) Financial resources
(c) Time resources
(d) Technological resources
(e) Mechanical resources
(f) Natural resources

1.4 WHAT IS PROJECT MANAGEMENT?

Project management is an application of knowledge, skills, tools and techniques to meet the project requirements. The objective is to use the optimum resources to accomplish a goal in minimum time.

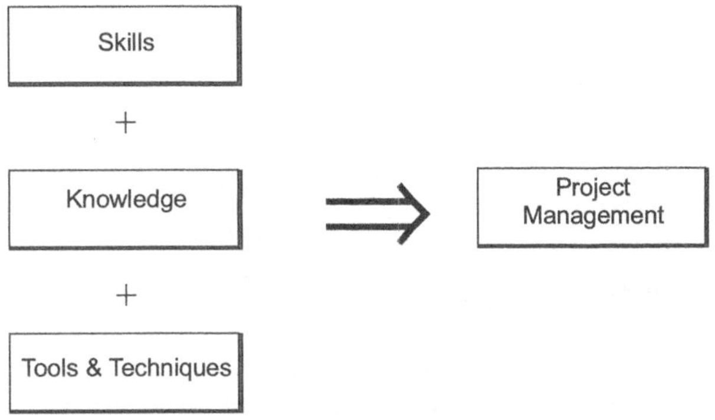

The 7-S of Project Management

The 7-S framework provides a comprehensive set of issues that need to be considered. It also allows classification of tasks within the remit of the project manager, which reduces the complexity of the role. In addition, classifying issues in this manner ensures that the project manager will know where to look to find sources of help if novel situation arises. Knowing that interpersonal problem in a team are aggravated by the style /culture that a project manager promotes provides a means for finding solutions to the problem

Strategy: The high-level requirements of the projects and the means to achieve them

Structure: The organizational arrangement that will be used to carry out the project

Systems: The methods for work to be designed monitored and controlled

Staff: The selection, recruitment, management and leadership of those working on the projects

Style / culture: The underlying way of working and inter-relating within the work team or organization

Stakeholders: Individuals and groups who have an interest in the project process or outcome

1.5 CHARACTERISTICS OF PROJECT MANAGEMENT

1. **Project Charter:** The project charter is one of the special project management documents that translates the project sponsor's business case into project objectives.
2. **Tools and Techniques:** The management of projects uses a number of special tools and techniques, combined with subject related knowledge and skills that have been developed over the years to manage different aspects of the project and facilitate the processing of large amounts of data. These include the CPM analysis, the Gantt chart, the resource histogram, the earned value graph and tables, activity crashing and matrix organization structures.
3. **Project Plan:** The project plan is a special project management document integrating the knowledge areas with individual plans to form one combined baseline plan.
4. **Project Management Processes:** The project management process is a special management technique for managing a linear sequence of steps or interrelated actions performed to achieve a specified set of projects, results or services. The project management process is subdivided into the following sub processes (initiation, planning, execution and closing).
5. **Project Organization Structure:** The project organization structure is a special management technique that enables the project manager to form temporary organization structures and project teams that can be designed to suit the needs of the project and the project participants. The enables the project leader to build and motivate the team and coordinate their work.
6. **Project Methodology:** The project lifecycle format is a special management technique subdividing the project into a number of identifiable phases that each produce a distinct deliverable. This sequence of phases forms the backbone of the project methodology, interlinking all the topics within a phase and between phases.

Do you know?: A product life cycle has five stages among which project management is done in development stage

Note : 1. A portfolio can contain programs or projects specifically to meet strategic objectives.
2. Program $\rightarrow$ Group of related Projects.

1.6 DIFFERENT TYPES OF MANAGEMENT

1. **General Management:** The successful project manager must also be competent in a wide range of general management skills. These include:
   - Recruiting and team selection
   - Sales and marketing
   - Project economics
   - Legal contracts
   - Personnel and human resources
   - Accounts and salaries
   - Computer systems
2. **Production Management:** Although projects are deemed to be unique, in reality they usually consist of a number of similar and repetitive tasks that could be grouped and made using a production line approach.
3. **Process Management:** Process management is used by companies that manage a product that typically flows from one process to another.
4. **Programme Management:** A programme is a set of related projects and organizational changes put in place to achieve a strategic goal and to deliver the benefits that the organization expects.
5. **Portfolio Management:** Portfolio management could be used by a project office that is running a number of unrelated projects such as managing the repairs and maintenance of a large telecom type company, a power station or a water utility.

A portfolio is a set of projects and/or programmes that are not necessarily related but are brought together for the sake of control, co-ordination and optimization.

Portfolio management is mainly applied to groups of projects and programmes that might not be related in the business sense but draw on a common pool of scarce resources. The portfolio manager co-ordinates all the projects and possible programmes within an organization during the process of evaluation, selection, monitoring and controlling, reprioritization and closure.

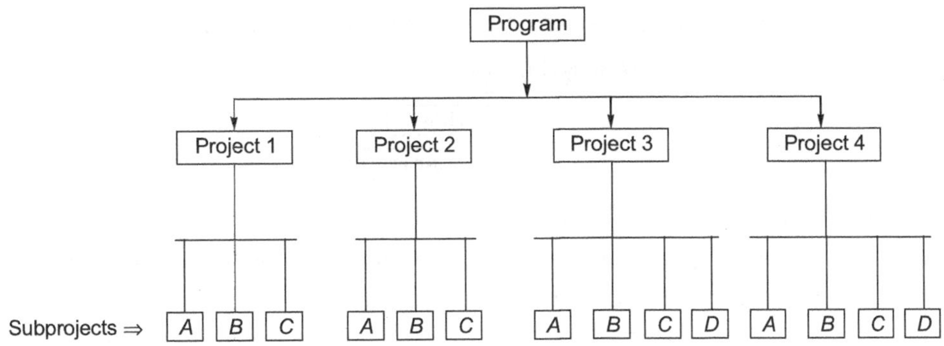

1.7 PROJECT LIFE CYCLE

- The project life cycle shows how a project can be subdivided into a number of phases presented sequentially along a project timeline.
- Every project has certain phases of development. A clear understanding of these phases allows managers and executives to control the project more efficiently.
- All the phases of a project from start to end are known as life cycle phases. The project life cycle i.e. number of phases may differ from project to project.

The main phases of life cycle are:

1.7.1 Project Life Cycle (4 Phases)

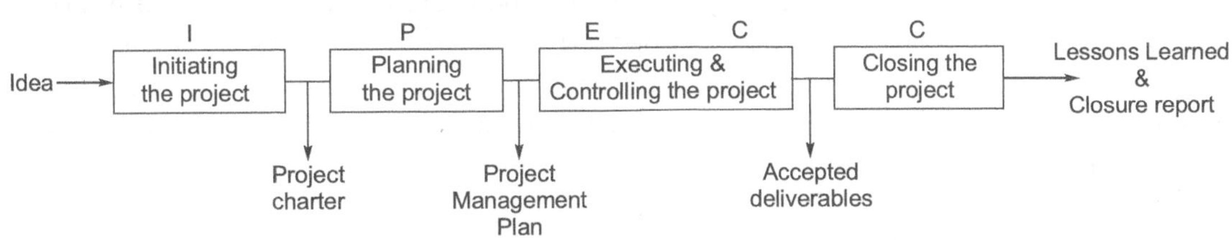

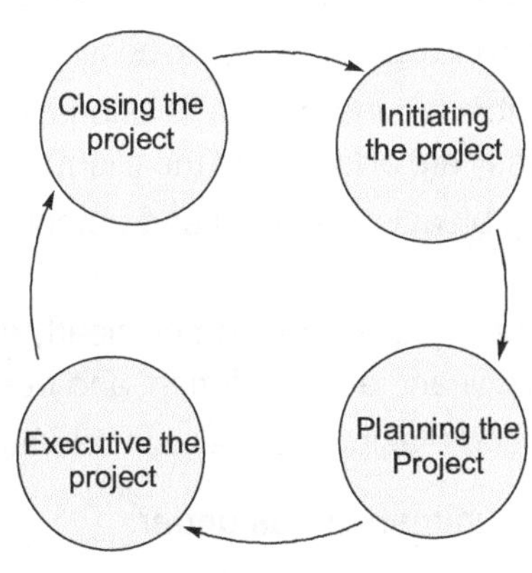

Project Life Cycle

Monitoring and control processes is an integral part of all the process groups, and interact with each process in numerous ways:

Type of Relation between Activities of a Project:

1. **Sequential :** Finish to start Relationship i.e. one phase starts when the preceeding has finished.

2. **Overlapping :** Here the subsequent phase can begin before the previous has finished.

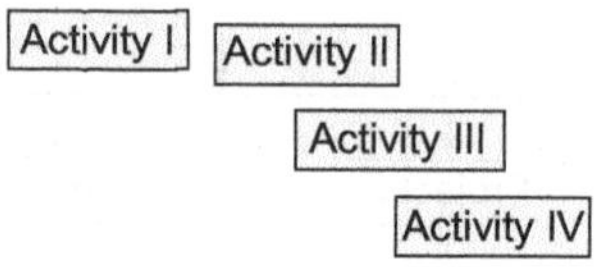

It saves time and generally done for IT Projects.

3. **Interative :** This phase type is particularly useful with Projects in which there are uncertainities rapidly changing market conditions or complete unknowns.

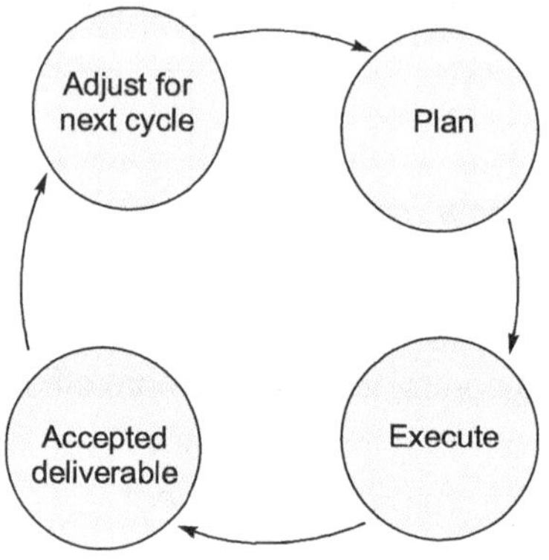

Concurrent Engineering: Concurrent engineering, or simultaneous engineering or Integrated Product Development or Agile, is a work methodology where the various areas required to bring a product from conceptualization to sales is done in parallel as opposed to step-by-step. The result is decreased time in product/service realisation/completion leading to increased time to market resulting in more sales and early market capture. Said areas could include research and development, design, manufacturing, testing, etc. for the product or project.

Phase-1. Feasibility Stage or Conception Stage/ Initiation stage

- The feasibility phase assesses the business case to confirm it is feasible to manufacture and implement.
- If there are number of possible business case solutions the feasibility study will rank the business cases in order of how well they are addressing the client's requirements.
- This is the phase when a problem is identified and potential solutions are suggested i.e. ideas are conceived.
- After feasibility study once the objectives have been clearly defined then the appraisal of the solutions is conducted in terms of risk, financial commitment and benefits.
- In case project ideas are found feasible from all considerations it is then given ***'go ahead'*** signal.

Phase-I of the project can be summarized as under:

(a) Identification of the need

(b) Establishment of the primary feasibility of project idea

(c) Identification of alternatives
(d) Evaluation / appraisal of the alternatives
(e) Investment decision

Phase-2. Design Stage (Planning & Scheduling Stage)/Definition Phase

- The project definition phase uses the guidelines from the feasibility study to design the product, outline the build-method and develop detailed schedules and plans (baseline plan) for all the knowledge area topics required to make the project.
- Once the investment decision is taken, the design or the planning stage of project starts.
- In this phase original ideas are amplified to prepare ***'blue print'*** for next stage. It means technical parameters are frozen and basic designing is completed and specification for equipments are finalized, costs are estimated in details, a time schedule for the project is planned and steps are taken for raising funds and resources at the end of the design phase blue print is ready for execution. The output of design phase is called ***"Detailed Project Report (DPR)"***.
- Usually DPR is further examined by the concern organisation. From first phase to second phase of the project life cycle, ***the intensity of activities continuously increases.***

Phase-3. Execution or Production Phase

- The project execution phase uses the design and project plan from the definition phase, together with the execution strategy, to construct the project.
- In third phase project moves for execution or production where the emphasis is given to give physical shape to the ideas presented in DPR.
- In this phase procurement of resources (material/machinery) starts.
- **The intensity of activities further builds up and reaches to peak in 3rd phase, however when execution approaches to completion the intensity of activities start falling again.** This is most important phase. The demands on the project manager is at its peak in this phase.
- There is a great need of continuous monitoring and control to all activities in this phase.

Phase-4. Termination or Commissioning or Handover Phase

- The project commissioning and handover phase inspects and confirms the project has been made to the approved design and then hands over the project to the client for operation.
- It is the last phase of the project cycle. During this phase the constructed facilities are tested one by one and final teething problems are solved. If trial is successful then the commissioning is complete.
- After commissioning, the project is handed over. This stage might include training of operating personals. In this phase ***intensity of activities reduces*** to minimal at the end.

Life cycle graph between intensity of activities and time:

Phase-I $\rightarrow$ Feasibility/Appraisal/ Conception
Phase-II $\rightarrow$ Design/Planning and scheduling / Development/ Definition
Phase-III $\rightarrow$ Implementation / Execution/Production
Phase-IV $\rightarrow$ Commissioning/Termination/ Transfer

The important elements of a project life cycle are:

1. Operations /activities; which should be performed in sequence.
2. Resources: manpower, material, money machinery etc.
3. Constraints and external conditions.

1.8 PROJECT CONSTRAINTS

With any project, there are limitations and risks that need to be taken into account and addressed to ensure the project's ultimate success. The three primary constraints that project managers should be familiar with are time, scope and cost. These are frequently known as the triple constraints or the project management triangle.Each constraint is connected to the other two; so, for example, increasing the scope of the project will likely require more time and money, while speeding up the timeline for the project may cut costs, but also diminish the scope.

There are four main parameters of a project management

1. **Time :** Time is very crucial to any project. Alteration of Project completion time changes the cost.
2. **Scope :** Scope tells what needs to be achieved and the work that must be done to deliver a project.
3. **Cost :** Cost is the monetary value of a Project. Cost is directly dependent on time and scope and the quality to be produced.
4. **Quality :** This is the standard of something as measured against other things of a similar kind, the degree of excellence of something.

The Triple Constraint: Here any change in one Parameter will most likely affect others. All three constraints of scope, cost and time are interrelated and changing any one will result in changing the other two also. The resultant of the three is quality.

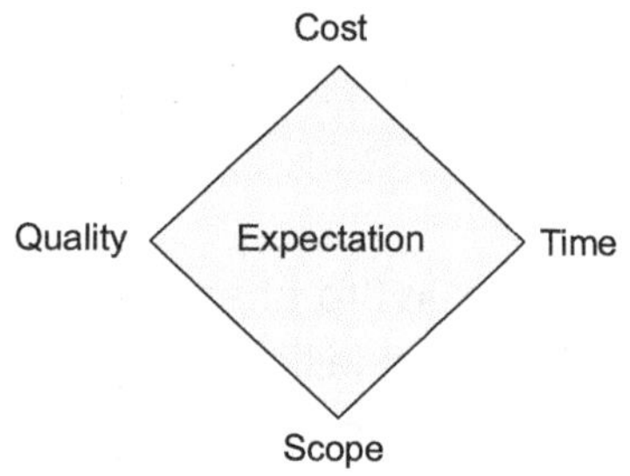

The Diamond Constraint: The Project management diamond constraint include cost, quality, time and scope.

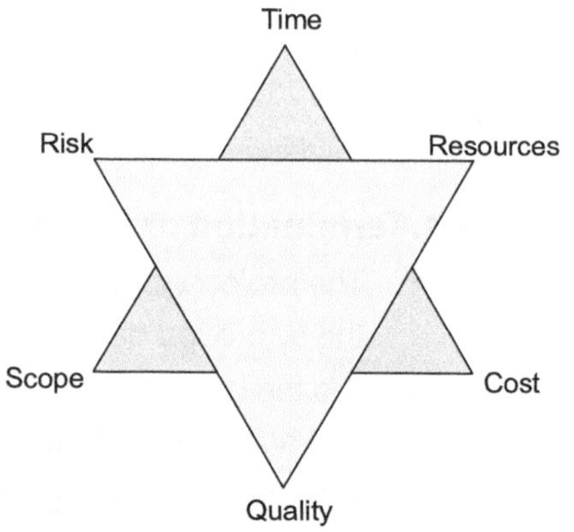

The Sextuple Constraint: Now-a-days apart from time cost and scope risk quality and resources are also constant for better understanding.

A risk is an event that may or may not happen. If it does happen, it will have unwanted consequences and will result in losses.

Here the change in any constraint will affect the other in order to maintain the expectation diamond. A constraint limits the execution of project or process.

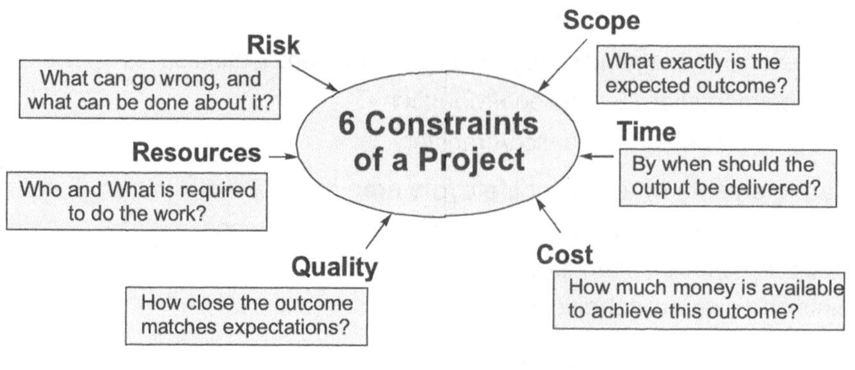

1.9 LEVEL OF EFFORT (INTENSITY OF ACTIVITIES)

The project lifecycle is often presented with its associated level of effort. The level of effort could be any parameter that flows through the project that can be measured, but it is most commonly expressed as man-hours or expenses/costs. The level of effort is a useful indicator for the project manager to quantify the amount of work to be performed and the amount of work completed within each phase. These parameters can be presented as a line graph of '*rate of expenditure*' (or rate of effort) and/or a line graph of '*cumulative expenditure*'.

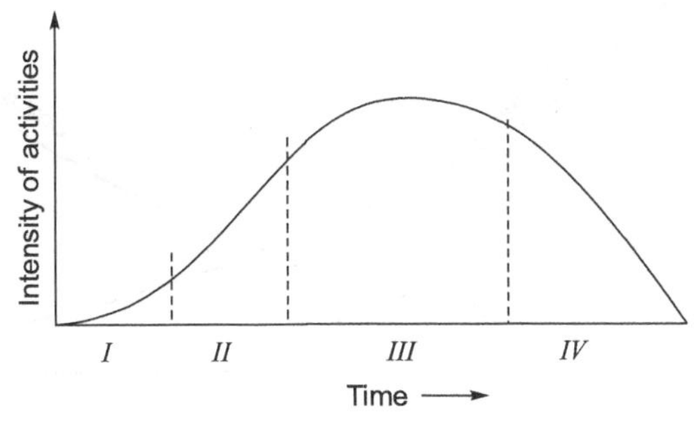

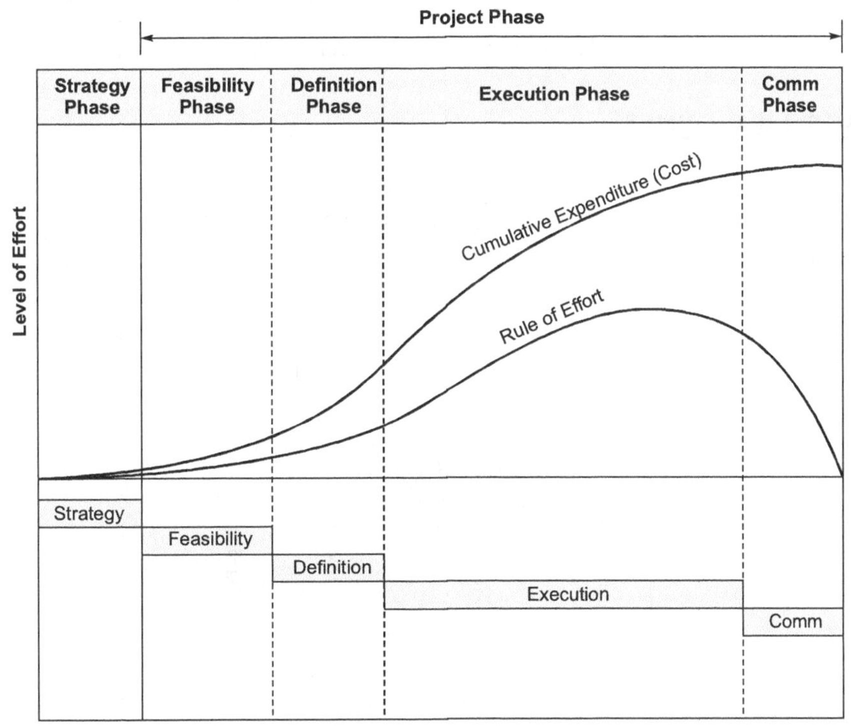

Fig. *Level of Effort*

The accumulated effort is the sum of the effort to date. This shows a typical 'S curve' profile similar to that used in the earned value calculation. This is a useful feature for the project manager to note as similar projects tend to have similar levels of effort profiles.

1.10 LEVEL OF INFLUENCE VS. COST OF CHANGES (FRONT-END IMPORTANCE)

'Cost of change' curve plotted against the project lifecycle.

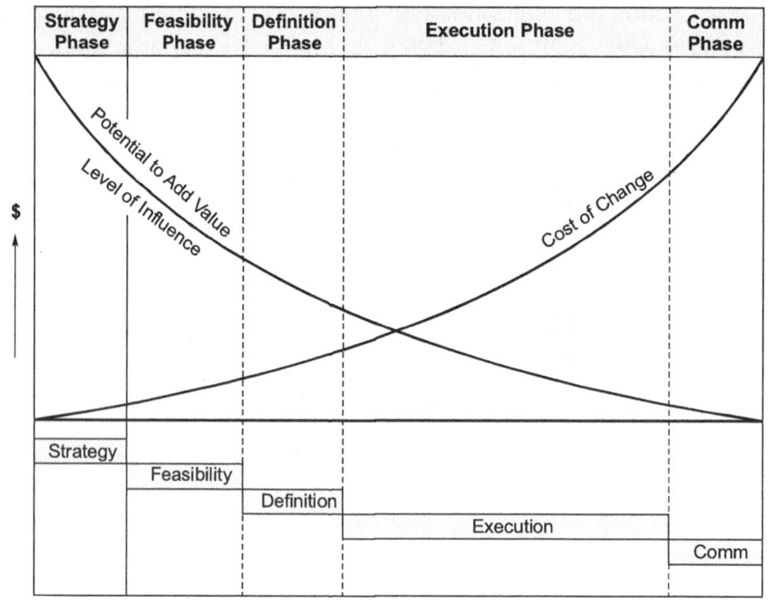

Fig. *Level of influence vs. Cost of change*

1.11 PRODUCT LIFE CYCLE

Products, like people, have life cycles. The product life cycle is broken into four stages: introduction, growth, maturity, and decline. This concept is used by management and by marketing professionals as a factor in deciding when it is appropriate to increase advertising, reduce prices, expand to new markets, or redesign packaging.

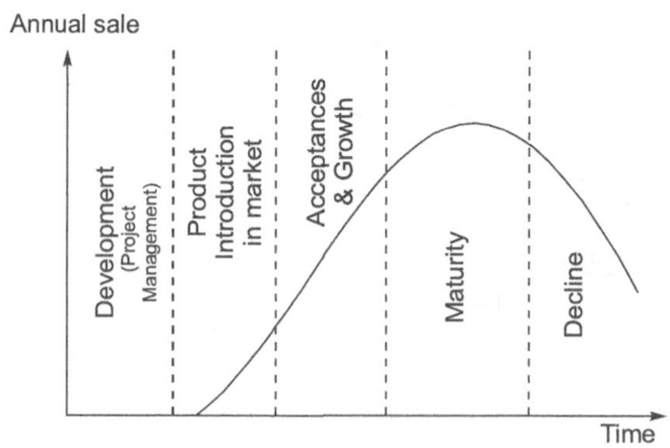

Fig. *Product Life Cycle*

What do the PLC stages mean?

Introduction: Introducing a new product where it's unknown and products are small. The price is often higher as distribution is limited and promotion is personalized.

Growth: Popularity for the product grown, meaning it is being bought in greater numbers and, with volume, the price declines. Distribution increases and promotion focuses on product benefits.

Maturity: The product competes with alternatives and its pricing drops. Distribution becomes intense (it's available everywhere) and promotion focuses on the differences to competitors' products.

Decline: Reaching the end of its life, the product faces fewer competitors. The price may rise and distribution has become selective as some distributors have dropped the product. Promotion aims to remind customers of its existence.

1.12 PRODUCT LIFE CYCLE (10 PHASES)

The classic product life cycle only considers the project from the feasibility study to the commissioning and handover (four phase). However, if the project is considered from the client's (project sponsor's) perspective, then the strategy phases and the operational phases need also to be included, as on the next page.

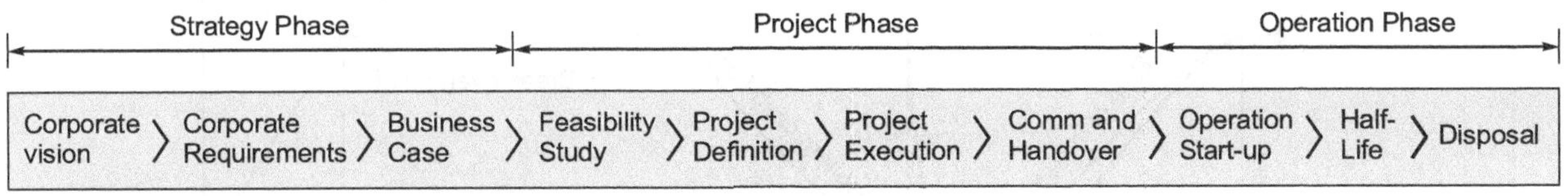

A. Strategy Phases

1. **Corporate vision and value phase:** The corporate vision and values phase establishes the corporate vision and values which outline the purpose and long term aims of the company, together with details of the company's culture, philosophy and the way the company intends to do business.
2. **Corporate requirement phase:** The corporate requirement phase investigates what the company needs to do to maintain competitive advantage to stay in business, and what opportunities the company could exploit to help achieve the long term corporate objectives.
3. **Business case phase:** The business case phase outlines the corporate strategy indicating how to solve the problems, requirements and opportunities with a number of proposals. The business case seeks to justify the use of company resources to pursue each course of action.

B. Project Phases

4. **Project feasibility phase or Initiation phase:** The feasibility phase assesses the business case to confirm it is feasible to manufacture and implement. If there are a number of possible business case solutions, the feasibility study ranks the business cases in order of how well they are addressing the client's requirements.
5. **Project definition phase:** The project definition phase uses the guidelines from the feasibility study to design the project, outline the build-method and develop detailed schedules and plans (baseline plan) for all the knowledge area topics that are required to make the project.
6. **Project execution phase:** The project execution phase uses the design and project plan from the definition phase to construct the project. An execution strategy is developed to make the best use of corporate resources and outsourcing.
7. **Project commissioning and handover or closing phase:** The project commissioning and handover phase inspects and confirms the project has been made to the approved design, then hands over the project to the client for operation.

C. Operation Phases

8. **Operation start-up phase:** The operation start-up phase implements the new facility, product or service into the operational environment. It is the project sponsor's responsibility to ensure the operation of the project realizes benefits for the client organization. From the client's perspective this is the main purpose of the project.

9. **Project upgrade phase:** The half-life upgrade phase incorporates the latest technology and systems to keep the project running efficiently and competitively.
10. **Project disposal phase:** The disposal phase brings the project to a formal closure by dismantling the facility and restoring the environment to its original state.

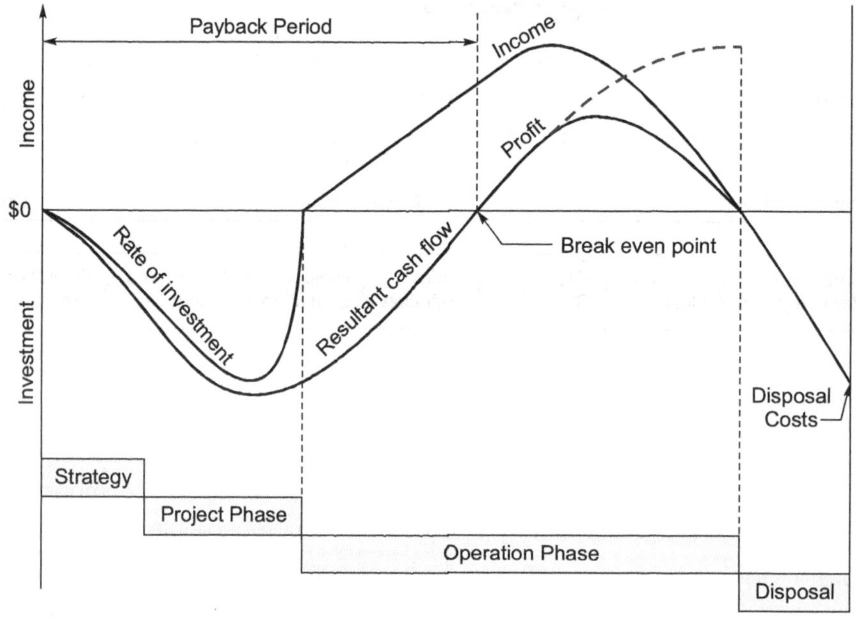

Fig. *Project lifecycle Costing*

1.13 PROJECT MANAGEMENT PROCESS

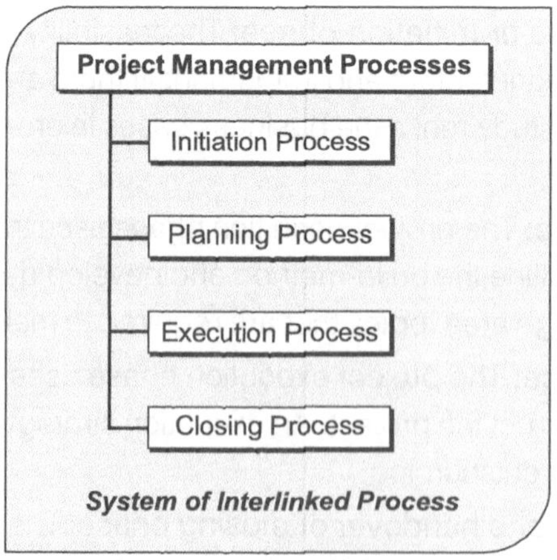

Fig. Systems approach

1. Initiation Process-Initiate the Phase

The initiation process is the first process within each phase that formally starts the phase. The project methodology systems approach subdivides the initiation process of each phase into the following sub processes:

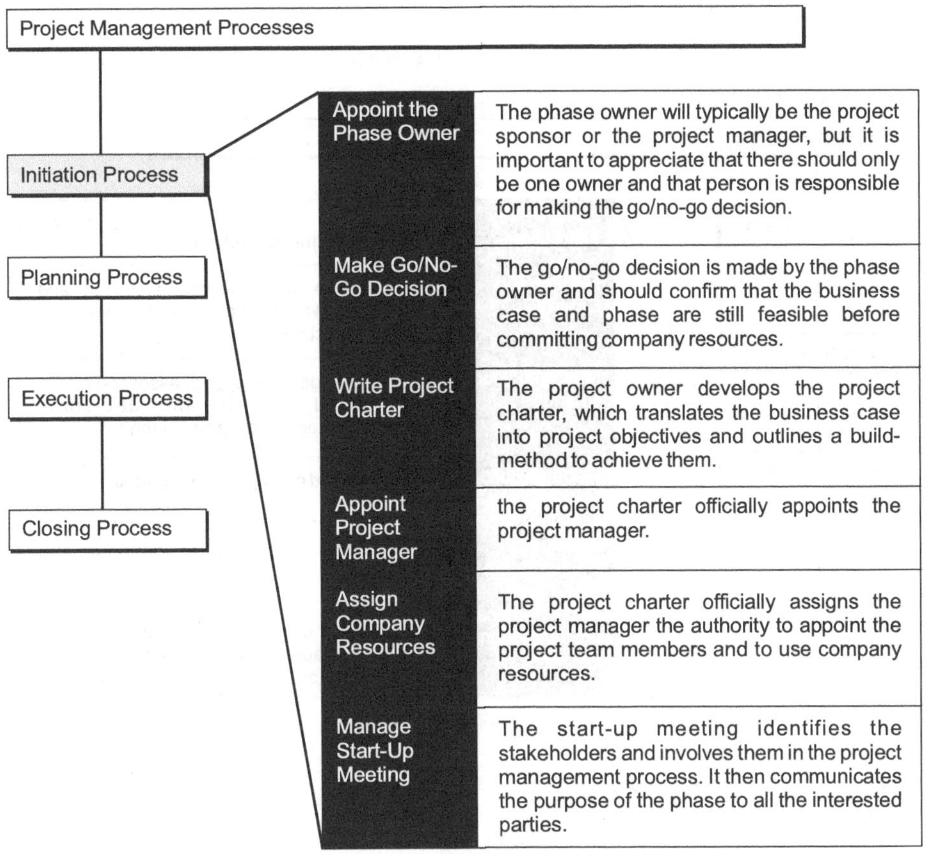

Fig. *Initiation process*

2. Planning Process-Plan and Schedule

The planning process is the second process within each phase which develops a fully integrated plan for the phase. The project methodology systems approach subdivides the planning process into the following sub processes:

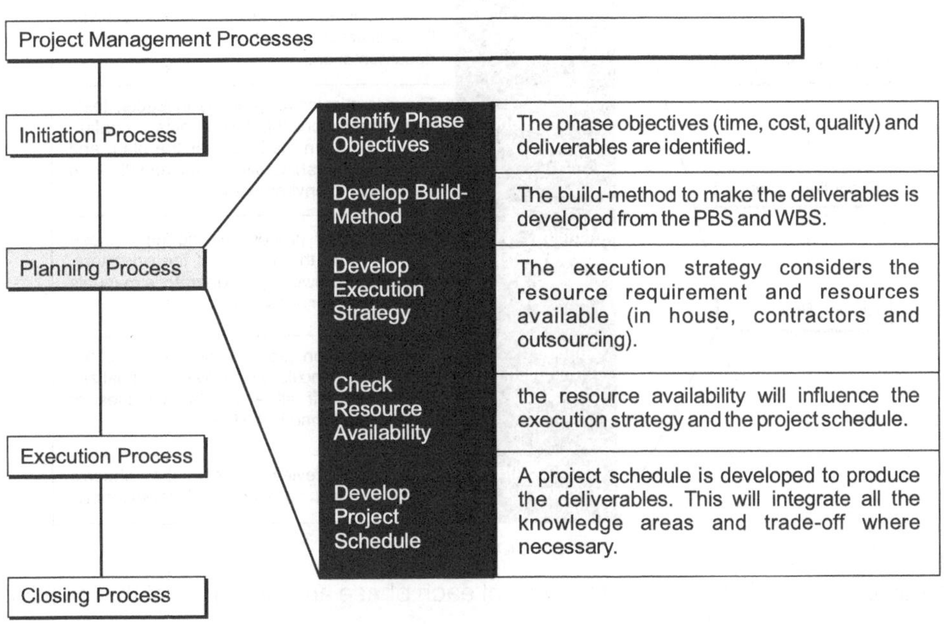

Fig. *Planning process*

3. Execution Process-Execute the Phase

The execution process is the third process within each phase that manufactures the phase deliverables. The project methodology subdivides the execution process into the following sub processes:

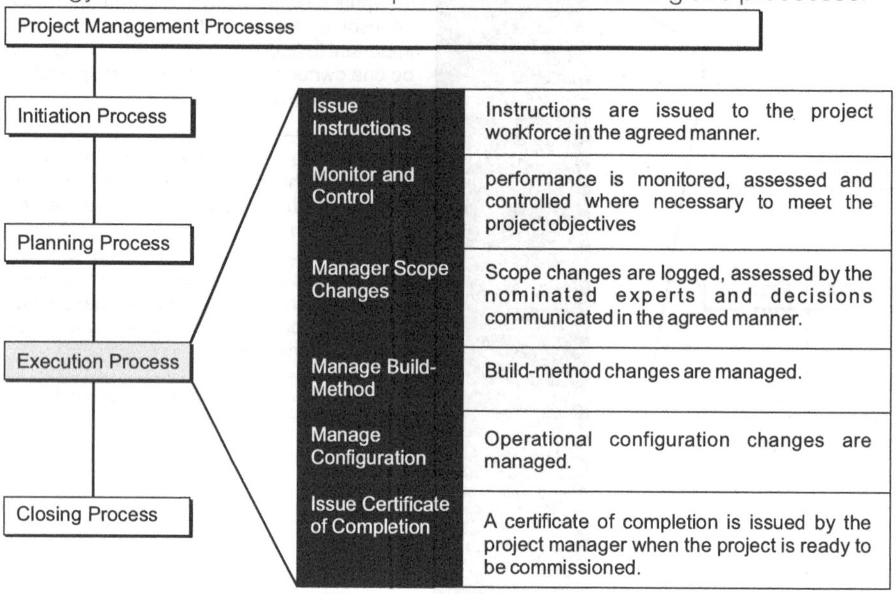

Fig. *Execution process*

4. Closing Process-Close the Phase

The closing process is the fourth process within each phase that accepts and confirms that the deliverables have been manufactured to the required condition and then hands them over to the next phase. The project methodology systems approach subdivides the closing process into the following sub processes:

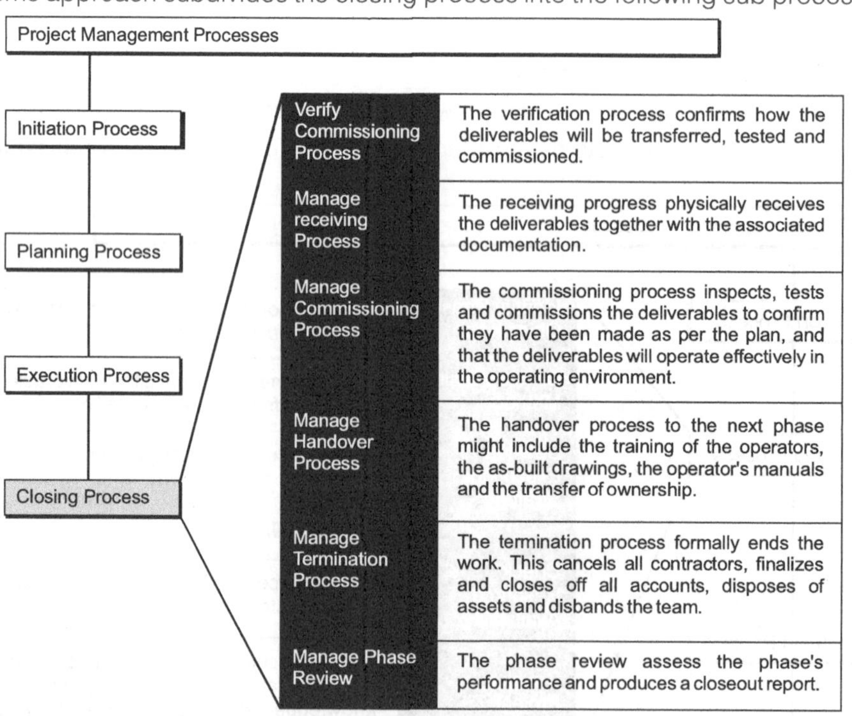

Fig. *Closing process*

A go/no-go decision can be made at the end of each phase and/or as part of the initiation process of the following phase.

Previous ESE Prelims Questions

Q.1 Consider the following statements regarding project management:

1. It is the process of attaining project objectives in a stipulated time to produce quantified and qualified deliverables
2. It is the art of bringing together the responsibilities, resources and people necessary to accomplish the business goals and objectives within the specified time limitations and within the financial grant

Which of the above statements is/are correct?

(a) 1 only (b) 2 only
(c) Both 1 and 2 (d) Neither 1 nor 2 [ESE Prelims : 2017]

Ans. (c)

Q.2 Consider the following phases of project management:

1. Identification 2. Formulation
3. Appraisal 4. Implementation

Which of the above phases are relevant, sequentially?

(a) 1, 2 and 3 only (b) 1, 2 and 4 only
(c) 3 and 4 only (d) 1, 2, 3 and 4 [ESE Prelims : 2017]

Ans. (d)

Q.3 **Statement (I):** Project management is essentially the process to plan its implementation and to pre-determine the period-wise need of resources including funds and personnel, given the choice of total duration and quality standards.

Statement (II): Of the four dimensions (not denying that there can be some more) of a project. viz., scope, cost, time and quality, only any two can be pre-assigned; others have to abide by these two prescriptions.

(a) Both Statement (I) and Statement (II) are individually true and Statement (II) is the correct explanation of Statement (I).
(b) Both Statement (I) and Statement (II) are individually true but Statement (II) is **not** the correct explanation of Statement (I).
(c) Statement (I) is true but Statement (II) is false.
(d) Statement (I) is false but Statement (II) is true. [ESE Prelims : 2018]

Ans. (a)

Q.4 In progress of a project, the percentage of error will be less in

(a) Definitive cost estimate (b) Detailed estimate
(c) Preliminary estimate (d) Study estimate [ESE Prelims : 2019]

Ans. (b)

Q.5 In a project life cycle, the maximum percentage of effort is done in

(a) Concept phase (b) Definition phase
(c) Planning and organizing phase (d) Implementation phase [ESE Prelims : 2019]

Ans. (d)

Q.6 **Statement (I)** : All projects have constraints or limitations that inhibit their ability to reach goals and objectives.

Statement (II) : Time and money are universal constraints in projects.

(a) Both Statement (I) and Statement (II) are individually true; and Statement (II) is the correct explanation of Statement (I)

(b) Both Statement (I) and Statement (II) are individually true; but Statement (II) is NOT the correct explanation of Statement (I)
(c) Statement (I) is true; but Statement (II) is false
(d) Statement (I) is false; but Statement (II) is true [ESE Prelims : 2019]

Ans. (a)

Q.54 According to Black's Law Dictionary, the law treats the corporation itself as a person which can
(a) monitor
(b) maintain
(c) sue and be sued
(d) give response [ESE Prelims : 2024]

Ans. (c)

Q.55 Those who drive automobiles are familiar with blind spots. Applying this term to organizational and business arenas, Dennis Moberg draws an analogy between business blind spots and those we experience when driving. Blind Spot is one of the significant common impediments to responsibility. Which one of the following is NOT the method under Blind Spot?
(a) Self-deception
(b) Willful blindness
(c) Inattentional blindness
(d) Illusion of invulnerability of group [ESE Prelims : 2024]

Ans. (a)

Q.56 Which one of the following is NOT a factor for large scale diversification into unrelated areas by some of the industry Conglomerate in India?
(a) Restriction in growth in the existing line of business
(b) Policies with respect to imports, duties, pricing and reservations
(c) Opening up of newer areas of investments
(d) Desire not to avail tax incentives [ESE Prelims : 2024]

Ans. (a)

Objective Brain Teasers

Q.1 The steps or phases taken by the project from beginning to ending are collectively called its:
(a) Hierarchy of steps to be followed on the project
(b) The manager's method of completing the project
(c) The project Life Cycle
(d) A collaborative approach between the Manager and the team to complete the work on the project

Q.2 The project management process groups are:
(a) Initiating, planning, expediting, and control.
(b) Plan, organize, develop, and control.
(c) Plan, do, observe, commit.
(d) Initiating, planning, executing, monitoring, controlling, and closing.

Q.3 Ideally, a project manager should be selected and assigned at which point in the project life cycle?
(a) During the initiating process or conception stage
(b) During the project planning process
(c) At the end of the design phase of the project life cycle
(d) Prior to the beginning of the design phase of the project life cycle

Q.4 The second largest employment generating industry in the country is
(a) Agriculture
(b) Petro-chemical
(c) Construction Industry
(d) Leather Industry

Q.5 ______ recognizes that a project or phase should begin and commits the organization to do so:
(a) Initiating Process
(b) Solicitation Process
(c) Scoping Process
(d) Planning process

Q.6 Projects are initiated in the life of an organization as a result of:
(a) An immediate financial need
(b) To keep employees busy and productive
(c) To achieve a pre-set strategic objective
(d) To make products for sale in the market

Q.7 A collection of generally sequential project phases whose name and number are determined by the control needs of the organization or organizations involved in the project, is called:
(a) Project life cycle
(b) Product life cycle
(c) Portfolio management
(d) Program management

Q.8 In which of the following project management process groups, most of the time and money are typically spent?
(a) Initiating (b) Planning
(c) Executing (d) Controlling

Q.9 All of the following are characteristics of a project except:
(a) Temporary
(b) Definite beginning and end
(c) Interrelated activities
(d) Repeats itself every month

Q.10 What do knowledge, skills, tools and techniques stand for when applied to the activities of a project, if it also entails effective management of a process.
(a) Human Resource Management
(b) Project Management
(c) Financial Management
(d) Project Management Process

Q.11 A project is defined as:
(a) A process of considerable scope that implements a plan.
(b) An endeavour, which is planned, executed, and controlled; performed by people; and constrained by limited resources.
(c) A temporary endeavour undertaken to create a unique product or service.
(d) An objective based effort of temporary nature.

Q.12 Which of the following statements is correct?
(a) Process Groups are same as Project Phases
(b) Each Process Group has a mapped Project Phase
(c) Process Groups are not Project Phases
(d) Monitoring and Control Process Group applies only in monitoring and control phase

Q.13 In case of a project being given by an external client to a company, the client must have provided all the following information except...
(a) Project Statement of Work
(b) Agreements
(c) Business Case
(d) Organizational Process Assets

Q.14 In general a project charter covers:
(a) Project justification, background, scope, objectives, available resources, a project schedule, estimates, procurement schedule, and any limitations that top management wants to apply.
(b) Project justification, background, scope, objectives, available resources, organizational responsibilities, estimates, and detailed procurement schedules responsibilities, and any limitations that top management wants to apply.
(c) Project justification, description, details of sponsor and project manager, budget, assumptions and constraints, boundaries to apply.
(d) Project scope, objectives, team responsibilities, a project schedule, estimates, procurement schedules, and any limitations that top management wants to apply.

Q.15 Output of a Project is
1. Product
2. Service
3. Improvement in Product
4. Continuing Mass Production
(a) All of these (b) 1, 2 and 3
(c) 1, 3 and 4 (d) 1, 2 only

Q.16 Project produces either service or product. It is also
1. Temporary 2. Unique
3. Operational in nature 4. Repeatable
(a) All of these (b) 1, 2 and 3
(c) 1 and 2 only (d) 1, 2 and 4

Q.17 Mass manufacturing of a bicycle in an industry is
(a) Project (b) Operation
(c) Both (a) and (b) (d) None of these

Q.18 The process in which application of knowledge skills, tools & techniques to project activities to meet project requirement is done.
(a) Process management.
(b) Programme Management
(c) Quality Management
(d) Project Management

Q.19 Stakeholder in a project can be
1. Customer 2. Project Team
3. Financer 4. Project Manager
(a) All of these (b) 1, 2 and 4
(c) 2, 3 and 4 (d) 2 and 4 only

Q.20 Write the Project Management Process in sequence
1. Initiation 2. Controlling & Monitoring
3. Executing 4. Planning
5. Closing
(a) 1, 2, 3, 4, 5 (b) 4, 1, 2, 4, 5
(c) 1, 4, 3, 2, 5 (d) 1, 4, 2, 3, 5

Q.21 A project
(a) Endures for the lifetime of the product, process or service
(b) has a "Progressive elaboration".
(c) has elements which are all predictable and repeatable
(d) does not meet strategic objectives of the project.

Q.22 A project charter is prepared and issued by
(a) Stakeholders
(b) Project manager
(c) Project sponsor
(d) Project management office

Q.23 In a project intensity of activities is maximum at
(a) Initiation phase (b) Planning Phase
(c) Execution Phase (d) Closing Phase

Q.24 Project charter is the outcome of
(a) Planning Phase (b) Initiation Phase
(c) Execution Phase (d) Closing Phase

Q.25 A project manager is made officially as project manager during
(a) Initiation
(b) Planning
(c) Before project phase start
(d) Any time in the between

Q.26 It is the first document of the project and authorises the project manager
(a) Risk register (b) Programme register
(c) Project charter (d) Logbook

Q.27 Elements in a project are all below given except
(a) continually evaluating risk
(b) definite start & end
(c) predictable & repeatable
(d) subjected to constraints like resource & schedule

Q.28 A stakeholder is affected by the project in
(a) Positive manner
(b) Negative manner
(c) Not affected
(d) May be positive or negative

Q.29 Project occurs in which phase of product life cycle
(a) Development (b) Product introduction
(c) Maturity (d) Decline

Q.30 Who among the following influence definitionof scope of the project?
1. Project Manger
2. Team member
3. Stakeholder
4. Customer
(a) 1, 2 and 3 only (b) 1, 2 and 4 only
(c) 2, 3 and 4 only (d) 1, 2, 3 and 4

Q.31 A company has combined its different projects to meet its strategic business objectives. This is called
(a) Project (b) Portfolio
(c) Programme (d) Crowdsourcing

Q.32 Which product life cycle phase gives the maximum sale?

(a) Product introduction
(b) Acceptance & Growth
(c) Maturity
(d) Decline

Q.33 Projects have some characteristics, which of the following relate their characteristics?
(i) They have a repetitive nature.
(ii) Risks of project failure is very high at start.
(iii) Stakeholder influence is most for product at the end of project.
(iv) Cost of project increase gradually at start and falls rapidly at end
(a) All of these (b) (i) and (iii) only
(b) (ii), (iii) and (iv) (d) (ii) and (iv) only

Q.34 Regarding the payback period in a project management which is true
(a) It is duration from start when rate of investment become positive
(b) It is duration from start when income starts from the project
(c) It is duration from start when the profit starts from the project
(d) It's duration is not related to breakeven point.

Q.35 Which of the following are in depended of project variation?
(a) Project life cycle phases
(b) Project management process groups
(c) Both (a) and (b)
(d) None of the above

Directions: Each of the next items consists of two statements, one labelled as 'Assertion (A)' and the other as 'Reason (R)'. Examine these two statements carefully and select the answers to these items using the codes given below :

Codes:
(a) Both Assertion (A) and Reason (R) are individually true and Reason (R) is the correct explanation of Assertion (A).
(b) Both Assertion (A) and Reason (R) are individually true but Reason (R) is NOT the correct explanation of Assertion (A).
(c) Assertion (A) is true but Reason (R) is false.
(d) Assertion (A) is false but Reason (R) is true.

Q.36 Assertion (A): Reading your favourite book is project.
Reason (R): Reading a book, which you have already read includes all predictable and repetitive element.

Q.37 Assertion (A): A project can have one or many phases.
Reason (R): Each phase of a project has IPECC process group.

Q.38 Assertion (A): A group of activity to produce a product or a service can be project or operation.
Reason (R): Point of view is very important to identity a project or an operation. As for customer the work may be project but for company doing similar activities it will be operation.

Q.39 Assertion (A): Detailed project report is prepared before planning starts.
Reason (R): Outcome of DPR will be used in execution of project activities.

Q.40 Assertion (A): A stakeholder is a project related person who is responsible for financing the Project only.
Reason (R): A stakeholder is anyone who is affected by the project.

Q.41 What is the reason for a Project Charter?
(a) To formally approve a venture or a stage and archive beginning prerequisites which fulfill the partner's needs and desires
(b) To archive how the venture will be arranged, executed, observed/controlled, and shut
(c) To connect the venture, which will be arranged, executed, and observed/controlled to the progressing work of the association
(d) To portray the way towards playing out the work characterized in the venture administration arrange so as to accomplish the venture's targets

Q.42 A project manager reviewing the scope of the project and the scope baseline of the project. This includes which of the following?
(a) The original project schedule, budget, and scope.
(b) The original project description and project charter.
(c) The original scope of the project plus or minus any scope changes.
(d) The current budget of the project.

Q.43 Arrange the following elements of the Project Cycle in the right order:

A. Project Appraisal B. Feasibility Analysis
C. Negotiation D. Project Selection

(a) A-B-C-D (b) B-A-C-D
(c) B-A-D-C (d) A-B-D-C

Q.44 Constraints common to projects include

(a) Scope, quality, schedule, budget, and risk
(b) Scope, teaming, planning, and resources
(c) Scope
(d) Resources and communication

Answers

1. (c)	2. (d)	3. (a)	4. (c)	5. (a)
6. (c)	7. (a)	8. (c)	9. (d)	10. (b)
11. (c)	12. (c)	13. (d)	14. (c)	15. (b)
16. (c)	17. (b)	18. (d)	19. (a)	20. (c)
21. (b)	22. (c)	23. (c)	24. (b)	25. (a)
26. (c)	27. (c)	28. (d)	29. (a)	30. (d)
31. (b)	32. (c)	33. (d)	34. (c)	35. (b)
36. (d)	37. (b)	38. (a)	39. (d)	40. (d)
41. (a)	42. (c)	43. (c)	44. (a)	

Explanations

3. (a)
When the project manager is selected and assigned to the project during initiation, several of the usual start-up tasks for a project are simplify (d) In addition, becoming involved with project activities from the beginning helps the project manager to understand where the project fits within the organisation in terms of its priority relative to other projects and the ongoing work of the organisation.

5. (a)
Creation of project charter is the specific activity that is done in the Initiating Process.

21. (b)
Example (a, c) belong to operation

25. (a)
Project charter officially authorises project manager.

26. (c)
Project charter the 1st document of project authorises project manager.

27. (c)
A project is neither predictable nor repeatable.

28. (d)
Stakeholder is anyone affected by the project either positively or negatively.

30. (d)
Stakeholders influence is most at the beginning of project.

32. (c)
Maturity phase gives the mad amount of product sale.

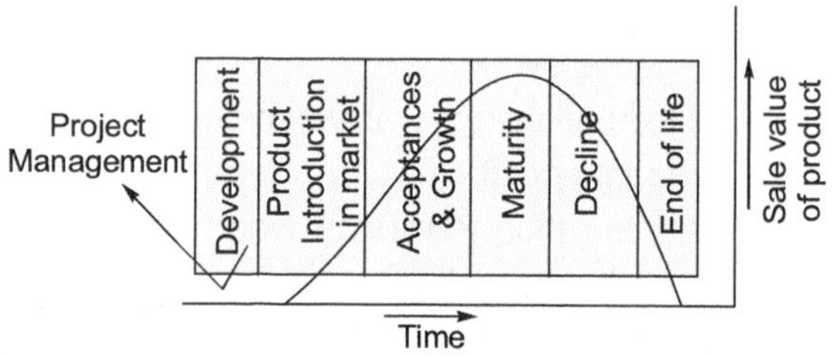

34. (c)
Payback period is when the breakeven point is reached and profit starts for the project.

35. (b)
Project management process groups i.e. IPECC remain same for all the projects whereas project phases may change depending on type of activity required
Example : Construction Project Phases will be different from software development project phase.

37. (b)
A project is broken down into number of phases which may be from one to many. But each phase includes IPECC cycle.

39. (d)
DPR is prepared in planning stage.

40. (d)
A is wrong. The world only is wrong there. Financer is one of the stakeholder.

44. (a)
The constraints include scope, schedule, budget (cost), quality, resources, and risk.

2 Project Organisation

2.1 PROJECT MANAGER

A project manager is a person who has the overall responsibility for the success of the project.The project manager also oversees any associated products and services, project tools and techniques to help ensure good practices. In addition, project managers are responsible for recruiting and building project teams, and making projections about the project's risks and uncertainties.

Project Manager must be versed in various functional areas like design, human resource, planning etc.

2.2 ROLE OF THE PROJECT MANAGER

Appointment: The project manager is appointed by the project sponsor as the single point of responsibility for completing the project a outlined in the project charter.

Management Skill: The project manager will require a range of management skills to plan and control all aspects to the project.

Leadership Skills: The project manager will require appropriate leadership skills and persuasive powers to lead, influence and negotiate with the project team, the contractors, the suppliers, the consultants and the stakeholders.

Project Team: The project manager is responsible for selecting and leading the project team.

Project Organization Structure: The project manager is responsible for setting up the temporary project organisation structure to integrate and interlink the resource provides (functional departments and contractors), suppliers and consultants.

Stakeholder Management: The project manager is responsible for identifying the project stakeholders and interested parties to determine and influence their needs and expectations. This process also engages the stakeholders to involve them in the decision making process.

Project Plan: The project manager is responsible for developing the project plan, which integrates all the individuals plans. The integration process has to make certain trade-offs and compromises between conflicting parameters so that the baseline plan can converge on an optimum arrangement.

Planning and Control System: the project manager is responsible for setting up and managing the planning and control system to issue instructions, monitor progress, review performance and take corrective action where necessary.

Scope Management: The project manager is responsible for identifying and defining the complete scope of work. The breakdown structure might use the PBS to identify the deliverables and the WBS to identify the work package to make the deliverables.

Scope Change Management: The project manager is responsible for managing the scope change process to assess the impact on the project's objectives and control any unnecessary scope creep.

Build-Method: The project manager is responsible for developing an efficient and feasible build method to develop the project.

Execution Strategy: The project manager is responsible for developing the executing strategy, which makes the 'make or buy' decision and selects the best option for the project.

Procurement Management: The project manager is responsible for setting up the procurement management system to procure all the required materials and components at the best price and at the right time. This process includes expending to make the procurement happen.

Resource Management: The project manager is responsible for developing the resource histogram and analysing the resource loading.

2.3 ATTRIBUTES/ QUALITIES OF A GOOD PROJECT MANAGER

An effective project manager is one who should have the following skills/capacities:

- Planning and organisational skills
- Personnel management skills
- Communication skills
- Ability to solve problems in their totality
- High energy levels
- Ambition of achievement
- Ability to take decisions
- Understanding the views of project team members and having a sympathetic attitude towards them
- Ability to develop alternative actions quickly
- Knowledge of project management methods and tools
- Ability to make self-evaluation
- Effective time management
- Capacity to relate current events to the project/project management
- Ability to handle project management software tools/packages
- Initiative and risk taking ability
- Familiarity with the organisation
- Conflict resolving capacity
- Team building skills
- Resource allocation skills
- Entrepreneurial skills
- Skills of negotiation

Power and its Use

Definition of Power: In political science or politics or business, Power can be defined as the ability of an individual to influence the actions of others to achieve a certain established goal. Legitimate power is called "authority" in organizations and is usually used by people in higher levels on their subordinates or team members to achieve the goals of the organization. The proper use of power does not involve the use of force or the threat of force (coercion).

Types of Power

There are five basic types of power:

- **Legitimate power, Formal power or Positional power:** This is the power invested in a person due to their position within the company or project. It is usually associated with the title, or chair, or uniform of the person and their duties and functions in the company. E.g.: the owner of the company.
- **Expert power:** This is the power gained by years of work experience or gaining of advanced educational qualifications in a particular area of work. Such people are called SMEs or subject matter experts. Their ability to understand and analyze problems and propose practical solutions to the problem gives them this power. This type of power is usually specific and limited to the area of expertise possessed by the SME. E.g.: Research scientists working at ISRO.
- **Referent power:** This is power gained by using interpersonal skills and personal charisma to attract and retain the loyalty of others and get them to think like you are thinking. E.g.: Thousands of people joined Mahatma Gandhi's Non Violent Movement because they were inspired by his personal traits and way of thinking and living.
- **Reward power:** This is the best kind of power to use if properly administered. Here a person uses power for the benefit of their employees or team members and gives them bonuses, rewards, new responsibilities, promotions, etc. for doing their job well.
- **Penalty or Coercive power:** This is the worst kind to use because here power is being used to penalize or deny rewards to employees for non-performance or disobedience. Though sometimes necessary, prolonged or continued use of only this kind of power results in a very negative work environment in the company.

Apart from these main types, other minor types of power that are used are: Professional power, Planning power, Risk power, Precedent power, ethical power, moral power, etc.

NOTE : Expert & Reward are best forms of power according to PMI. The worst is penalty.

Additional topics in Project Management

Strategy and types: Strategy can be defined as a high plan which includes contingency planning or options that accommodate for uncertainty in the original plan. It basically involves two main steps: Formulation or strategic thinking which looks at all options related to the situation at hand and picks the best one keeping in mind options required in uncertain situations. Implementation or execution is acting out the plan that has been selected for action.

As far as organizations are concerned, there are 3 types of strategy:

- **Business strategy** this defines the purpose and plan for the organization as a whole
- **Operational strategy** this defines what must be done on a daily basis to achieve the business goals
- **Transformation strategy** this involves radical and drastic changes to the business and operations strategies when the company decides to change its business strategy substantially

2.4 TEAMWORK IN PROJECTS

Team work can be defined as a sense or feeling of unity, enthusiasm, bonding and purpose developed among a group of persons closely associated in a task, project or work effort. It is a crucial necessity to achieve success in projects and programs. Whenever a group of people are put together, they go through the following steps, also known as the Tuckman Ladder. It is the responsibility of the Project Manager to ensure a quick and smooth transition between the first steps and get to the Perform stage at the earliest for project success.

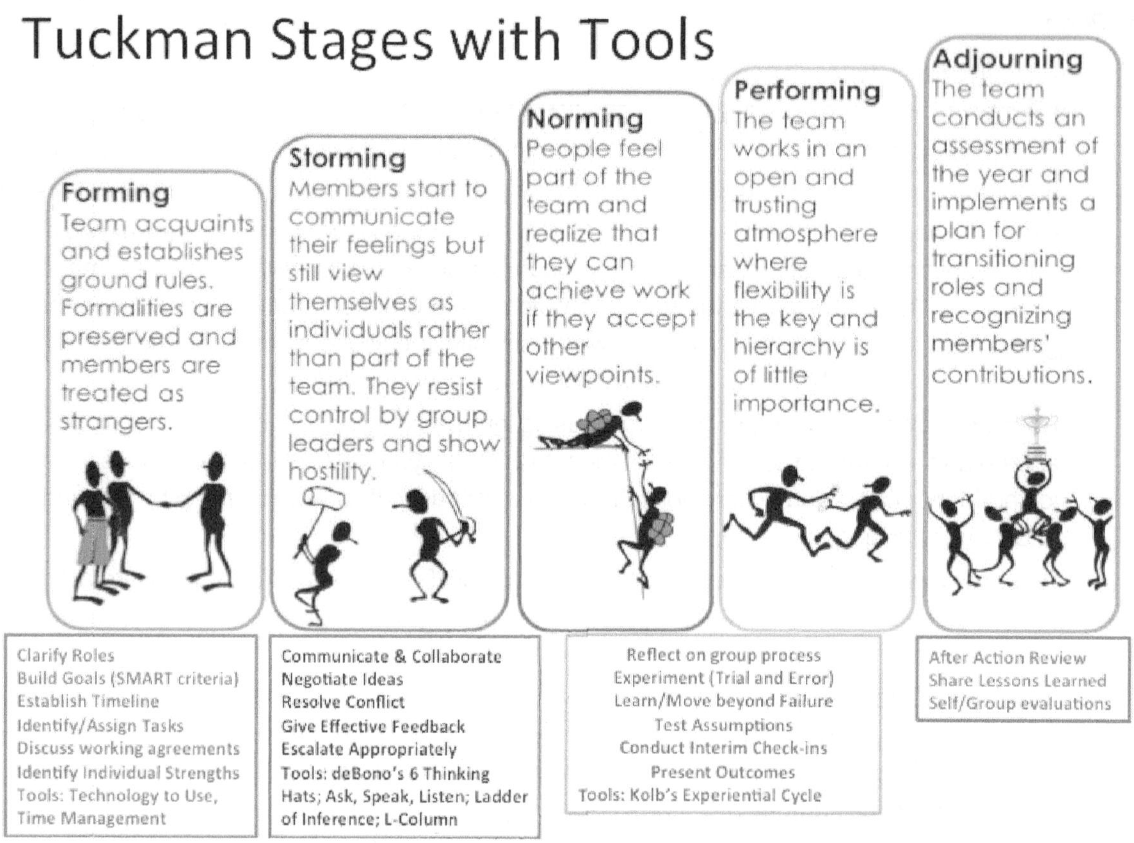

2.5 POWER/INTEREST GRID OF STAKEHOLDERS

Models to classify stakeholder ideas on basis of power/interest/influence:

1. **Power/interest :** Grouping stakeholders based on authority (Power) and interest
2. **Power/Influence :** Grouping stakeholders based on authority and influence (involvement)
3. **Influence/Impact :** Grouping stakeholders based on influence and project impact (ability to effect change)
4. **Salience Model :** Grouping based on power to influence, urgency (immediate need for attention) and legitimacy (appropriate involvement)

Among phase power /influence grid is shown below:

(a) If stakeholder has high power low influence – heap him satisfied
(b) If stakeholder is having high power high influence – manage him closely
(c) If stakeholder is having low power low influence – Just monitor him
(d) If stakeholder is having low-power high influence – Keep him informed.

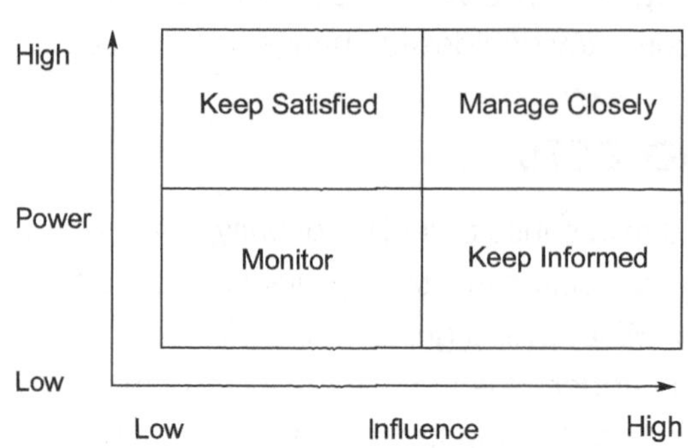

2.6 CONFLICTS MANAGEMENT BY PROJECT MANAGER

Conflict in project management is inevitable. The potential for conflict in information systems development projects is usually high because it involves individuals from different backgrounds and orientations working together to complete a complex task. The cause of conflict in team projects can be related to differences in values, attitudes, needs, expectations, perceptions, resources, and personalities. Proper skills in dealing with conflict can assist project managers and other organization members to handle and effectively resolve conflicts, which can lead to a more productive organization as a whole.

There has been a significant change in the manner conflicts are seen in a project. The difference is summarised as below:

Old View	New View
Conflict is inevitable	Conflict is necessary
It should be avoided	It allows team to grow
It is best resolved by avoidance or upper management intervention	It is Best resolved by the team member & near managers.

Causes of Conflicts:

1. Schedule
2. Resources
3. Technical opinion
4. Administrative overhead
5. Cost
6. Personality
7. Project Priority

Conflicts Resolution:

1. **Confrontation** – Facing the issue head to head from front.
2. **Collaboration** – In this two parties come to some viewpoint.
3. **Compromise** – Each side gives up something & everyone is dissatisfied, its a loose - loose situation.
4. **Forcing** – One party is forced by use of power.
5. **Smoothing** – Issue is not addressed directly, smoothens the conflicts by various ways such as agreement.
6. **Withdrawing** – Issue is avoided completely.

PMI declares confronting to be best followed by collaboration. The worst resolution technique is forcing any party.

DO YOU KNOW?

What is Halo effect?: Halo effect is the assumption that because the person is good at a technology/technical skills, then he will be a good project manager. It describes an error in thinking in which you make an inference about a person.

2.7 THE PMO

PMO is project management office which centralises the management of projects of the organisation.

Role of PMO :

1. Providing various methologies & polices for project management.
2. Providing guidance and training to project related individuals.
3. A resources pool of project managers for various organisational initiatives.
4. Manages dependencies between, projects, progress, or portfolios.
5. Functions as one of the stakeholder.
6. A centralised conduct for project.

Types of PMO

There are 3 types of Project Management offices in organizations as described by the Project Management Institute (PMI) in PMBOK 6:

1. **Supportive PMO:** Supportive PMO's act as a repository to your organization's projects. It supplies templates, best practices, training and keeps track of lessons learned. These have a very low degree of control in projects.
2. **Controlling PMO:** Controlling PMO's act as the auditor of the company. They check whether organizational tools, processes, and standards are applied in projects. They have a moderate degree of control in projects.
3. **Directive PMO:** As the name suggests, Directive PMO's manage and have a high degree of control in projects. Directive PMO's take complete control of all the projects executed within the organization.

2.8 TYPE OF PROJECT ORGANISATION STRUCTURE

Broadly speaking, organisational structure can be of three types. They are:

1. Functional organisation structure
2. Projectised organisation structure
3. Matrix organisation structure

Max. PM power over project

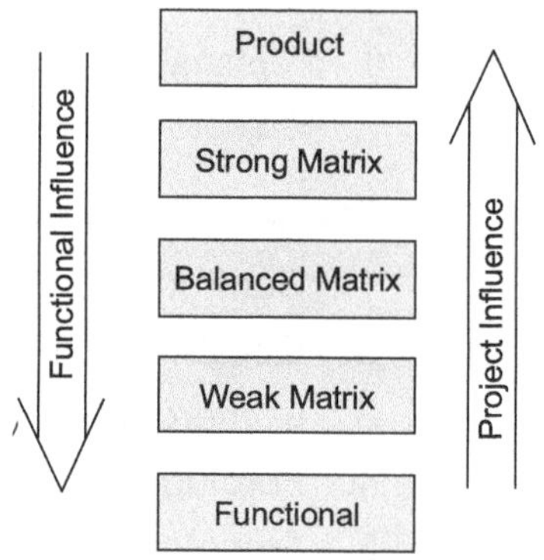

Min PM power over proejct

2.9 FUNCTIONAL ORGANISATION/TRADITIONAL/CLASSICAL ORGANISATION

- Functional organisation is the most basic and logical form of organisational structure. Functional organisation brings together in one department all those who are engaged in one activity or more related activities.
- It is most stable form of organisation.
- It has long line of communication and long chain of command.
- Project related issues are resolved by the concern functional head.

Example 1.

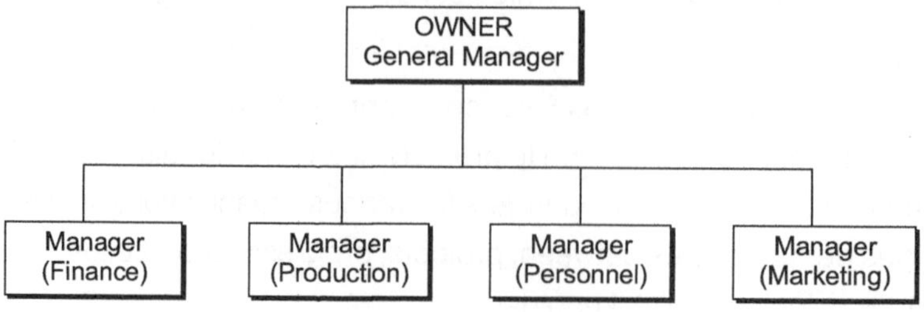

Fig. (a) *Functional organisation*

Example 2.

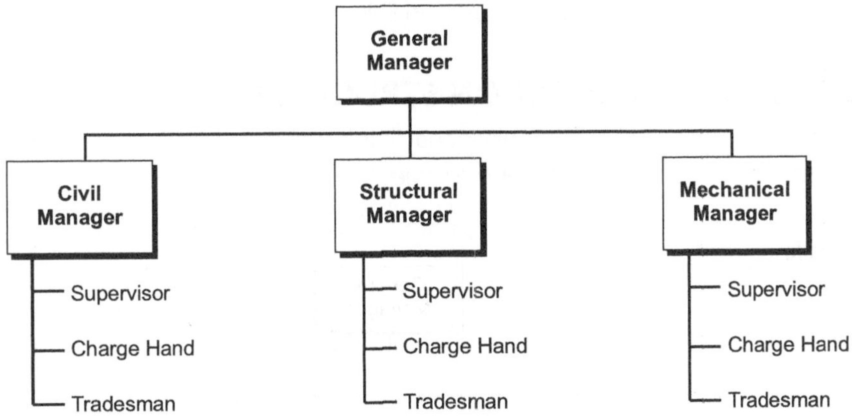

Fig. (b) *Functional Organisation Structure*

- This type of organisational structure is suitable for smaller organisations that offer a limited line of products. As the organisation grows by expanding its product line or when the organisation expands geographically., the weakness of functional organisational structure will become apparent since controlling will become ineffective. Another disadvantage of functional organisation is that it is difficult to fix accountability and difficult to judge the performance of the members.

Advantages:

- Better technical control is possible.
  - Specialists can be grouped to share knowledge and responsibility.
  - Personnel can be used on many different projects.
  - All projects will benefit from the most advanced technology (better utilization of scarce personnel).
- Flexibility in the use of manpower.
- A broad manpower base to work with/greater division of labour.
- Continuity in the functional disciplines; policies, procedures, and lines of responsibility are easily defined and understandable.
- Admits mass production activities within established specifications.
- Good control over personnel, since each employee has one and only one person to report to.
- Communication channels are vertical and well established.
- Quick reaction capability exists, but may be dependent upon the priorities of the functional managers.

Disadvantages:

- No one individual is directly responsible for the total project (i.e., no formal authority; committee solutions).
- Does not provide the project-oriented emphasis necessary to accomplish the project tasks.
- Coordination becomes complex, and additional lead time is required for approval of decisions.
- Decisions normally favour the strongest functional groups.
- No customer focal point.
- Response to customer needs is slow.
- Difficulty in pinpointing responsibility; this is the result of little or no direct project reporting, very little project-oriented planning, and no project authority.

- Motivation and innovation are decreased.
- Ideas tend to be functionally oriented with little regard for ongoing projects.

2.10 PROJECTISED ORGANISATION STRUCTURE

- The major advantage of this organisational flow is that one individual, the program manager, maintains complete line authority over the entire project.

Example:

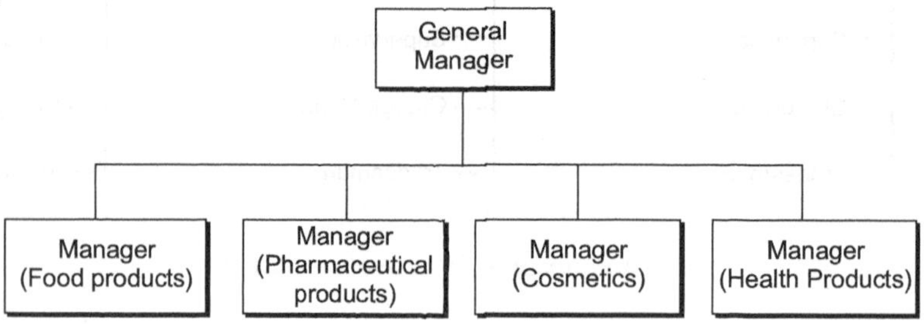

Fig. *Product organisation*

- In pure product organisations, long lead time becomes a thing of the past.
- In pure functional (traditional) structures, technologies are well developed, but project schedules often fall behind. In the pure project structure, the fast reaction time keeps activities on schedule, but technology suffers because without strong functional groups, which maintain interactive technical communication, the company's outlook for meeting the competition may be severely hampered. The engineering department for one project might not communicate with its counterpart on other project, resulting in duplication of efforts.

Advantages:

- Provides complete line authority over the project (i.e., strong control through a single project authority).
- Participants work directly for the project manager. Unprofitable product lines are easily identified and can be eliminated.
- Strong communications channels.
- Staffs can maintain expertise on a given project without sharing key personnel.
- Very rapid reaction time is provided.
- Personnel demonstrate loyalty to the project; better morale with product identification.
- A focal point develops for out-of-company customer relations.
- Flexibility in determining time (schedule), cost, and performance trade-offs.
- Interface management becomes easier as unit size is decreased.
- Upper-level management maintains more free time for executive decision-making.

Disadvantages:

- Cost of maintaining this form in a multi-product company would be prohibitive due to duplication of effort, facilities, and personnel; inefficient usage.
- A tendency to retain personnel on a project long after they are needed. Upper-level management must balance workloads as projects start up and are phased out.

- Technology suffers because, without strong functional groups, outlook of the future to improve company's capabilities for new programs would be hampered (i.e., no perpetuation of technology)
- Control of functional (i.e., organisational) specialists requires top-level coordination.
- Lack of opportunities for technical interchange between projects.
- Lack of career continuity and opportunities for project personnel.
- Resources sharing is minimal.

2.11 MATRIX ORGANISATION

- The matrix organisational form is an attempt to combine the advantages of the functional structure and the product organisational structure. This form is ideally suited for companies, such as construction, that are "project-driven".
- Each project manager reports directly to the vice president and general manager. Since each project represents a potential profit center, the power and authority used by the project manager come directly from the general manager. The project manager has total responsibility and accountability for project success.
- Project management is a "coordinative" function, whereas matrix management is a collaborative function division of project management.

Ground rules for matrix development:

- Participants must spend full time on the project.
- Horizontal as well as vertical channels must exist for making commitments.
- There must be quick and effective methods for conflict resolution.

Example 1.

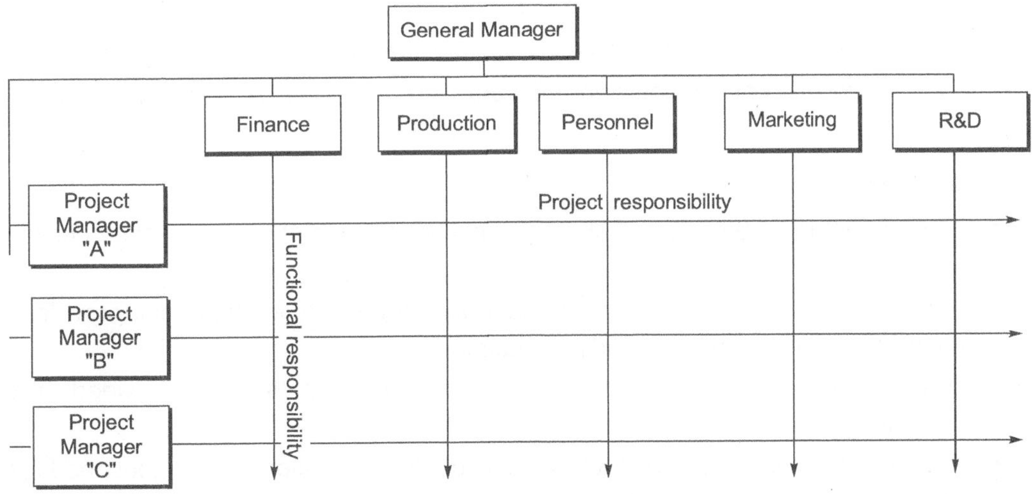

Fig. (a) *Matrix organisation*

Example 2.

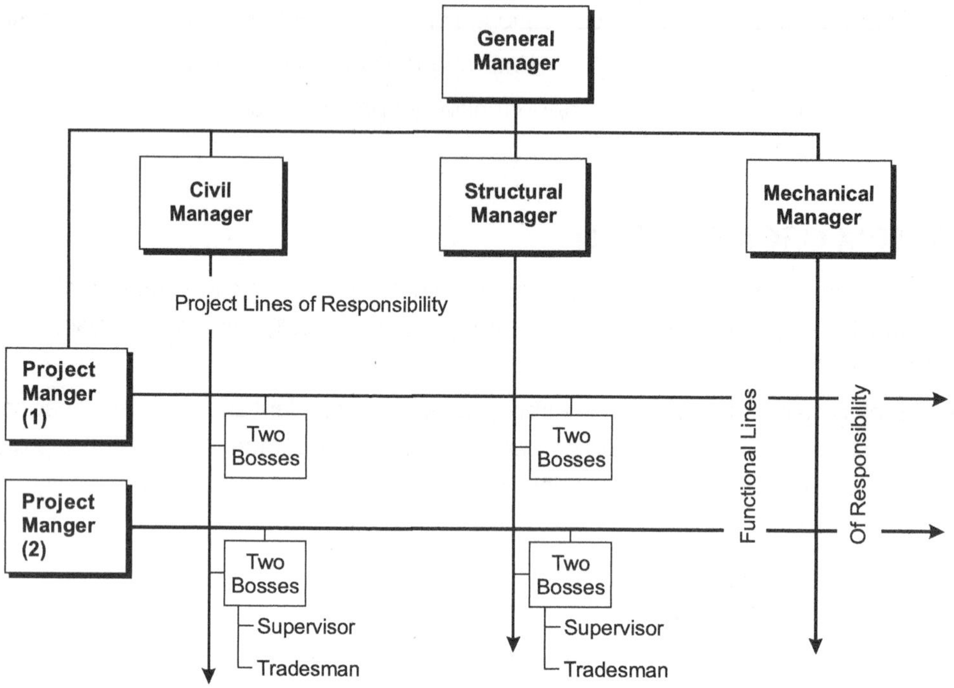

Fig. (b) *Matrix Organisation Structure*

- There must be good communication channels and free access between managers.
- All managers must have input into the planning process.
- Both horizontally and vertically oriented managers must be willing to negotiate for resources.
- The horizontal line must be permitted to operate as a separate entity except for administrative purpose.
- No two working environments are the same, and therefore, no two companies will have the same matrix design.

Advantages:

- The maximum project control (through the line managers) over all resources, including cost and personnel.
- Policies and procedures can be set up independently for each project, provided that they do not contradict company policies and procedures.
- The project manager has the authority to commit company resources, provided that scheduling does not cause conflicts with other projects.
- Rapid responses are possible to changes, conflict resolution, and project needs (as technology or schedule).
- The functional organisations exist primarily as support for the project.
- Each person has a "home" after project completion. People are susceptible to motivation and end-item identification. Each person can be shown a career path.
- Because key people can be shared, the program cost is minimized. People can work on a variety of problems; that is, better people control is possible.
- A strong technical base can be developed, and much more time can be devoted to complex problem-solving. Knowledge is available for all projects on an equal basis.

- Conflicts are minimal, and those requiring hierarchical referrals are more easily resolved.
- There is a better balance among time, cost and performance.
- Rapid development of specialists and generalists occurs.
- Authority and responsibility are shared.
- Stress is distributed among the team (and the functional manager).

Disadvantages:

- Multidimensional information flow.
- Multidimensional work flow.
- Dual reporting.
- Continuously changing priorities.
- Management goals different from project goals.
- Potential for continuous conflict and conflict resolution.
- Difficulty in monitoring and control.
- Company-wide, the organisational structure is not cost-effective because more people then necessary are required, primarily administrative.
- Each project organisation operates independently. Care must be taken that duplication of efforts does not occur.
- More effort and time are needed initially to define policies and procedures, compared to traditional form.
- Functional managers may be biased according to their own set of priorities.
- Balance of power between functional and project organisations must be watched.
- Balance of time, cost and performance must be monitored.
- Although rapid response time is possible for individual problem resolution, the reaction time can become quite slow.
- Employees and managers are more susceptible to role ambiguity than a traditional form.
- Conflicts and their resolution may be a continuous process (possibly requiring support of an organisational development specialist).
- People do not feel that they have any control over their own destiny when continuously reporting to multiple managers.

2.12 THE STRONG, WEAK OR BALANCED MATRIX

- Matrix structures can be strong, weak, or balanced. The strength of the matrix is based upon who has more influence over the daily performance of the workers: project manager or line managers. If the project manager has more influence over the worker, then the matrix structure functions as a strong matrix as seen through the eyes of the project manager. If the line manager has more influence than does the project manager, then the organisation functions as a weak matrix as seen by the project manager.
- The most common differentiator between a strong and weak matrix is where the command of technology resides: project manager or line managers. If the project manager has a command of technology and a recognized by the line managers and the workers as being a technical expert, then the line managers will allow the workers to take technical direction from the project manager. This will result in a strong matrix structure. Workers will seek solutions to their problems from the project manager first and the line managers second. The reverse is true for a weak matrix. Project managers in a strong matrix generally possess more authority than in a weak matrix.

- When a company desires a strong matrix, the project manager is generally promoted from within the organisation and may have had assignments in several line functions throughout the organisation. In a weak matrix, the company may hire from outside the organisation but should at least require that the person selected understand the technology and the industry.
- In weak matrix project managers authority is considerably reduced and it comes close to the functional structure of orgnisation.

2.13 PROJECT EXPEDITOR AND PROJECT COORDINATOR

In Weak Matrix organizations, there are usually no "real" project managers; even if there is a post of "Project Manager", they usually perform the role more of a Project Coordinator or Project Expeditor. For strong matrix and projectized organizations, Project Expeditor and Project Coordinator may exist alongside with the Project Manager.

Project Expeditor: A Project Expeditor usually carries the responsibility of staff assistant / communication coordinator.A Project Expeditor is not given the authority (or very low authority) to make or enforce decisions.The Project Expeditor will communicate with various parties of the project to ensure timely action.For larger projects, the Project Manager may have some Project Expeditors assisting them for communication and logistics.

Project Coordinator: A Project Coordinator carries the responsibility of partially managing the project under the supervision of other managers.A Project Coordinator is usually given some sort of limited authority to make decision.For larger projects, the Project Manager may have some Project Coordinators reporting to them.

2.14 PROJECT ORGANISATION STRUCTURE

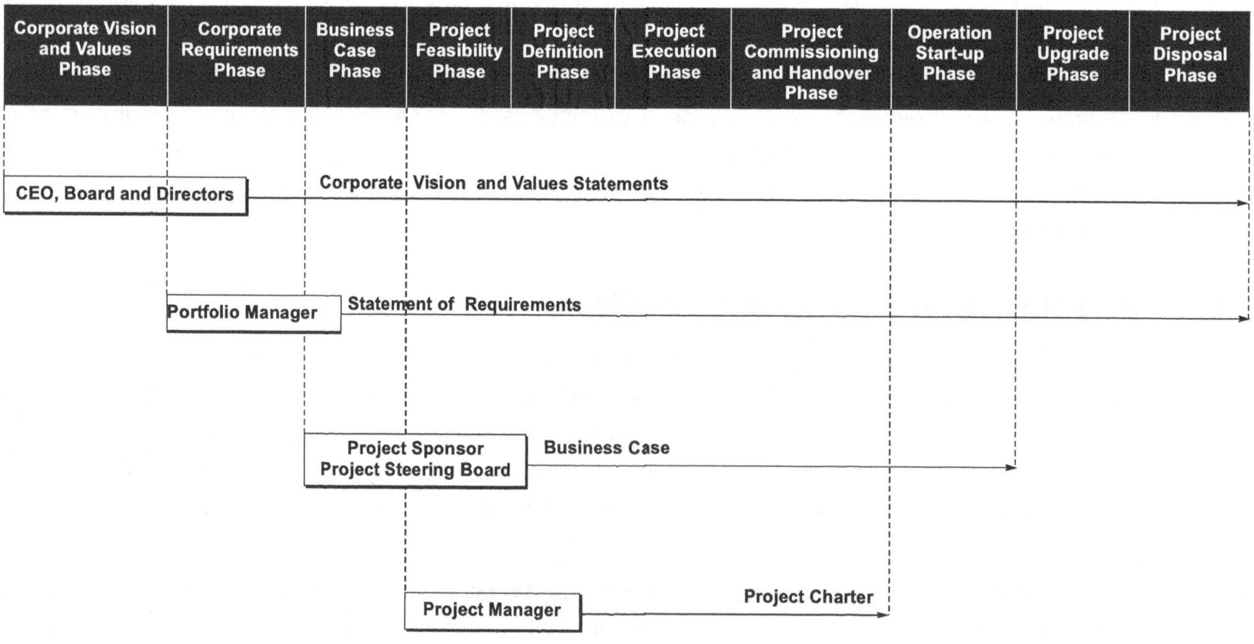

Fig. *Project Organisation Structure*

Previous ESE Prelims Questions

Q.1 Which term refers to a single person having authority to oversee all aspects of a product's production scheduling, inventory, dislocation and sales?

(a) Project management
(b) Product management
(c) Commercial management
(d) Venture management

[ESE-2019]

Ans. (b)

Q.2 Who is responsible for establishing, documenting and maintaining procedures for post-production handling functions such as storage, packaging and delivery?

(a) Production Manager
(b) Marketing Manager
(c) Vendor
(d) Quality Supervisor

[ESE-2019]

Ans. (c)

Q.3 Who is responsible for the following activities in a project?

1. Achieving a unity of control over project activities.
2. Having an authority to control project matters and disburse funds from the budget.
3. Having no actual line of authority over workers.

(a) Project Expeditor
(b) Project Coordinator
(c) Matrix Manager
(d) Project Manager

[ESE-2020]

Ans. (b)

Q.4 **Statement (I):** If the project influence is more in decision-making for the project, then the arrangement is considered a strong matrix.

Statement (II): If functional departments are seen to be influencing the decision-making more, the arrangement is considered a weak matrix.

(a) Both Statement (I) and Statement (II) are individually true and Statement (II) is the correct explanation of Statement (I)
(b) Both Statement (I) and Statement (II) are individually true but Statement (II) is not the correct explanation of Statement (I)
(c) Statement (I) is true but Statement (II) is false
(d) Statement (I) is false but Statement (II) is true

[ESE-2020]

Ans. (b)

Q.5 The Boston Consulting Group matrix classifies business in four categories as "STAR", "QUESTION MARK", "CASH COWS". Which one of the following is the fourth one?

(a) CATS
(b) HORSES
(c) DOGS
(d) HENS

[ESE-2021]

Ans. (c)

- The Boston Consulting Group (BCG) growth-share matrix is a tool used internally by management to assess the current state of value of a firm's units or product lines.
- The BCG growth-share matrix contains four distinct categories: "DOGS", "CASH COWS", "STAR", and "QUESTION MARK".

Q.6 Boston Consulting Group, the BCG matrix classifies the various businesses in a firm's portfolio on the basis of
(a) Relative Share and Relative Growth Rate
(b) Relative Market Share and Substantial Market Share
(c) Relative Market Share and Relative Market Growth Rate
(d) Substantial Growth Rate and Relative Market Growth Rate [ESE Prelims : 2024]

Ans. (c)

Objective Brain Teasers

Q.1 Match **List-I** with **List-II** and select the correct answer using the codes given below the lists:

List-I
A. Translates policy into a method of achieving the objective set out
B. Consists of defining the responsibilities of employees
C. Transmits all the information to the supervising staff
D. The organisational setup is aided to operate efficiently with flow of information, decisions and results in all directions

List-II
1. Co-ordinating
2. Planning
3. Organizing
4. Directing

Codes :

	A	B	C	D
(a)	3	2	1	4
(b)	2	3	1	4
(c)	3	2	4	1
(d)	2	3	4	1

Q.2 Match **List-I** (Indications of terms) with **List-II** (Terms) and select the correct answer using the codes given below the lists:

List-I
A. Used for recording instructions given by the Executive Engineer at site
B. Used widely for civil engineering construction
C. One of the principles of organisation
D. One of the functions of management

List-II
1. Co-ordination
2. Unity of command
3. Line organisation
4. Site order book

Codes:

	A	B	C	D
(a)	4	3	2	1
(b)	4	2	3	1
(c)	2	4	1	3
(d)	4	1	2	3

Q.3 The "halo effect" refers to the tendency to:
(a) promote from within
(b) hire the best
(c) move people into project management because they are good in their technical fields
(d) move people into project management because they have had project management training

Q.4 The overall co-ordinating person in a construction team is
(a) The Owner
(b) The Architect
(c) The Project Manager
(d) The Contractor

Q.5 The ability of a Project Manager or influence the costs on a project:
(a) Increases with time
(b) Remains the same over time
(c) Depends on the availability of resources and funds
(d) Decreases over time

Q.6 A type of organization in which the project manager has little formal authority and basically acts as a staff assistant to an executive who is ultimately responsible for the project is called:
(a) Balanced matrix (b) Weak matrix
(c) Projectized (d) Strong matrix

Q.7 Who is ultimately responsible for quality management on the project?
(a) Project engineer (b) Project manager
(c) Quality manager (d) Team member

Q.8 A project manager is trying to complete a software development project, but cannot get enough attention for the project. Resources are focused on completing process-related work and the project manager has little authority to properly assign resources. What form of organization must the project manager be working in?
(a) Functional (b) Matrix
(c) Expediter (d) Coordinator

Q.9 What are the major advantages of the functional type of organization?
(a) Single point of contact for the customer
(b) Stable organizational structure
(c) Project orientation
(d) Multifunctional teams are easy to form

Q.10 In which organization form would the project manager possess the greatest amount of authority?
(a) Classical/traditional
(b) Projectized
(c) Strong matrix
(d) Weak matrix

Q.11 In which organizational form do project managers have the greatest likelihood of possessing reward power and have a wage-and-salary administration function? (The project and line manager are the same person.)
(a) Classical/traditional
(b) Projectized
(c) Strong matrix
(d) Weak matrix

Q.12 In which type of matrix structure would a project manager most likely have a command of technology?
(a) Strong matrix (b) Balanced matrix
(c) Weak matrix (d) Cross-cultural matrix

Q.13 During project staffing, the primary role of line management is
(a) Approving the selection of the project manager
(b) Approving the selection of assistant project managers
(c) Assigning functional resources based upon who is available
(d) Assigning functional resources based upon availability and the skills set needed

Q.14 A project manager joining a new project should have following qualities
1. Leadership skill
2. Political knowledge
3. Communication skill
4. Project team confidence
(a) All of the above (b) 1, 2 and 3
(c) 1, 3 and 4 (d) 1 and 2 only

Q.15 An organisation where decision making authority is not with project manager is called
(a) Strong matrix
(b) Product matrix
(c) Weak matrix
(d) Balanced matrix

Q.16 The best forms of power of PM according to PMI are:
1. Expert 2. Legitimate
3. Reward 4. Penalty
(a) All of the above (b) 1 and 2
(c) 1 and 3 (d) 3 and 4

Q.17 You are working in an organisation which is having different department for different work function each headed by a manger. Such that total production of any project requires consents of these various department.
Which type of organisation it is?
(a) Projectized (b) work
(c) Functional (d) matrix

Q.18 Who is responsible for training of team members in a project?
(a) Project manager
(b) stakeholders

(c) they themselves
(d) project management office

Q.19 Project manager controls the project budget and having a full time role. Which type of organisation is it
(a) weak organisation
(b) balanced organisation
(c) functional organisation
(d) strong organisation

Q.20 Various stages followed in the formation of a project team in proper order are
1. Norming 2. Storming
3. Forming 4. Adjourning
5. Performing
(a) 1, 2, 3, 4, 5 (b) 3, 1, 2, 5, 4
(c) 3, 2, 1, 5, 4 (d) 1, 3, 2, 5, 4

Q.21 In a project management the project performance appraisal addresses the performance of the
(a) Team as a whole
(b) Individuals on the team
(c) Project manager
(d) All stakeholders

Q.22 The best way to resolve a conflicts in a project is by
(a) collaborative (b) confronting
(c) compromising (d) smoothing

Q.23 Which of the following skills of the project manager helps in conflict resolution?
1. Leadership
2. Technical
3. Influencing
4. Effective decision making
(a) All of the above (b) 1, 2 and 3
(c) 1, 3 and 4 (d) 1 and 3 only

Q.24 Which of the following type of organisation is more prone to the conflicts?
(a) strong matrix
(b) product organisation
(c) balanced organisation
(d) weak matrix

Q.25 What are triple constraint of a project?
(a) Time – quality – cost
(b) Cost – scope – time
(c) Time – quality – scope
(d) Scope – quality – cost

Q.26 What are the diamond constraints of a project?
(a) Time – cost – quality – expectation
(b) Time – cost – scope – quality
(c) Scope – cost – quality – expectation
(d) Scope – time – quality – expectation

Q.27 Which of the following may give rise to a conflict in a project?
1. Conflict over schedule of project
2. Conflict over resource sharing
3. Conflict over leadership roles
4. Conflict due to presentability of one's own
(a) All of the above
(b) 1, 2 and 3
(c) 2, 3 and 4
(d) 2 and 3 only

Q.28 At what stage of team formation, the team can work effectively even without a leader?
(a) Forming (b) Storming
(c) Norming (d) Performing

Q.29 Which of the below is correct with regard of project management?
(i) Project manager has max power over project in product organisation.
(ii) In weak matrix project manager has most control over funds.
(iii) In function project organisation people don't generally loose their job after project completion
(a) All of the above
(b) (i) only
(c) (i) and (iii)
(d) (ii) and (iii)

Q.30 In project management type of matrix organisation are :
1. Functional 2. Weak
3. Balanced 4. Strong
5. Product
(a) All of the above (b) 2, 3 and 4
(c) 1, 2, 4 and 5 (d) 2, 4 and 5

Q.31 The method used to collect information to use through all phases of the project life cycle is called:

(a) Responsibility matrix
(b) Organization breakdown structure
(c) Work breakdown structure
(d) Priority matrix

Q.32 Effective project team should have?
A. Clear sense of mission
B. Enough Cash
C. Cohesiveness
D. Trust
(a) A, B, C (b) A, C, D
(c) All of the above (d) A only

Answers

1. (d)	2. (a)	3. (c)	4. (c)	5. (d)
6. (b)	7. (b)	8. (a)	9. (b)	10. (b)
11. (a)	12. (a)	13. (d)	14. (b)	15. (c)
16. (c)	17. (c)	18. (d)	19. (d)	20. (c)
21. (b)	22. (b)	23. (c)	24. (d)	25. (b)
26. (b)	27. (a)	28. (d)	29. (c)	30. (b)
31. (c)	32. (b)			

Explanations

3. (c)
Flaw in Expert Judgment when people good in certain areas are deemed as good in all areas.

14. (b)
A project manager might be new to a project team and a complete stranger. So it is not necessary that team will have confidence in project manager.

18. (d)
In strong matrix authority and involvement of project manager is very high.

19. (d)
Forming – Storming – Norming – Performing – Adjourning

24. (d)
In weak matrix authority of project manager is less & several other functional manager have authorities increasing chances of conflict.

25. (b)

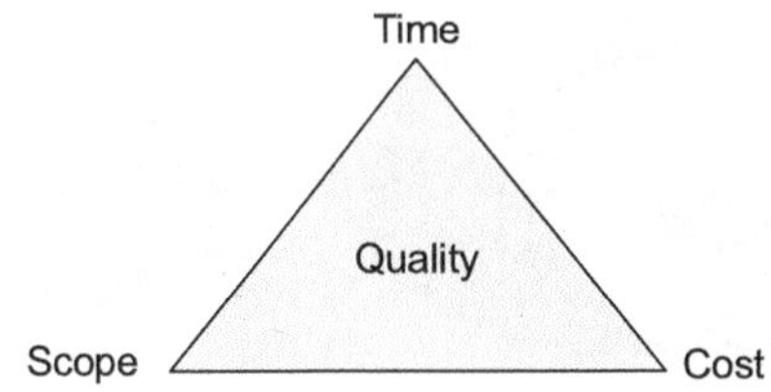

26. (b)

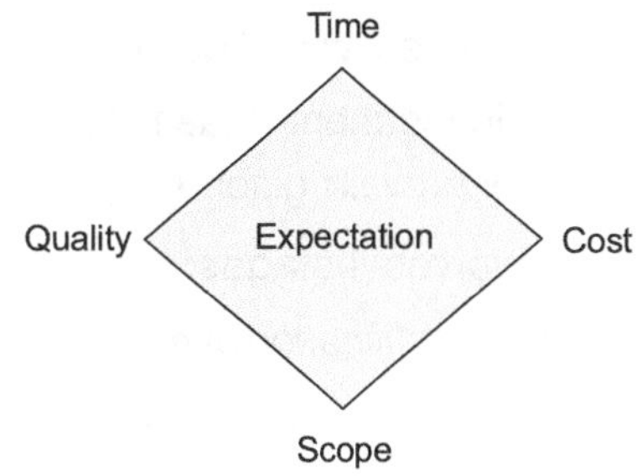

28. (d)
At performing stage the team is strategically aware of the work and situations around. Team understand their goals clearly and they can fulfill their goals even without a leader.

31. (c)
The work breakdown structure approach allows us to visually see the work that is needed in order to complete a project. The bottom line is by using a work breakdown structure it reduces the number of surprises and improves the ability to better estimate future projects. The work breakdown structure can be used to effectively decompose the project scope, to improve estimating, to better control the project execution and to more accurately verify project completion. In addition, using a work breakdown structure approach summarizes project information to improve the opportunity for use of historical information, which, can aid in both speed and accuracy of future projects. The work breakdown structure is a reportable process that can be used as a template for future projects.

■■■■

3 INITIATION PROCESS : Project Formulation and Appraisal

A typical project would go through a development process having following three phases.

1. **Pre-investment Phase:** It is related to authorization of investment decision for particular project idea under prevailing conditions.

   It involves 4 stages:

   (a) *Identification of relevant project idea*

   (b) *Preliminary filtration of project ideas through pre-feasibility studies*

   (c) *Detailed feasibility studies*

   (d) *Final evaluation and decision*

2. **Investment Phase:** It involves detailed design and actual implementation.

   This phase involves several inter-disciplinary tasks and has the following 4 phases:

   (a) *Negotiating and contracting*

   (b) *Detailed project design and engineering:* This stage covers site investigation, design and approval of lay-out, approval of drawings and blue prints preparation of time schedule and PERT charts, final selection of technology and equipments and detailed estimation of costs.

   (c) *Construction and Erection (Execution and control)*

   (d) *Trial runs, commissioning and optimization.*

3. **Operation Phase:** It is post commissioning stage when project is run for output or benefits.

DO YOU KNOW?

Magnitude of the risk can be reduced considerably by following a rational procedure in which:

(a) a sufficiently large number of good-quality ***alternatives****, whether mutually exclusive or otherwise, is generated for consideration.*

(b) reliable and adequate ***information*** *is gathered about each alternative including its sub-options, and*

(c) the sub-options and the alternatives are (in that order) subjected to systematic and objective ***evaluation*** *in relation to technical, economic, financial and other pertinent parameters for selection of the best course of action.*

3.1 PRE-INVESTMENT PHASE

1. Identification of Project Idea/Conceptualisation of Project Idea/Opportunity Studies

No investor, whether an individual or an organisation, would like to miss out on profitable opportunities either to enter a new attractive business, or to expand an existing business.

By their very nature, opportunity studies are indicative rather than detailed and hence are generally based on macro-parameters and rough estimates. Such studies may also be carried out at a more general level to identify opportunities in one of the following three directions.

(a) a given geographical area (Area studies)

(b) a specific sub-sector of economy or industry, like food-processing, pharmaceuticals or power generation (Sub-sectorial studies.)

(c) exploitation of renewable and nonrenewable natural, agricultural, or industrial produce like minerals, sugarcane, flash, plants, and even geothermal energy sources etc. (Resource-based studies).

Identifying a new project is a complex problem. It involves careful study from many different angles. Following sources can be used to get ideas for new projects.

- Performance of existing industry
- Import-export statistics
- Data from various sources
- Plan outlays and govt. plans
- Analysis of economic & social trends
- Availability of raw material
- Price trend
- Research result
- Identifying unfulfilled needs of market
- Review of sick units

Following Techniques are used while Project Identification

1. SWOT analysis

   S – Strength

   W – Weakness

   O – Opportunity

   T – Threat

   This is a good technique which considers internal and external factors and helps in Project Identification.

	Positive	Negative
Internal factors	**Strength (S)** 1. Technical know how within 2. Established organisation and structure 3. Collaborations	**Weakness (W)** 1. Lack of skilled persons. 2. Lack of Proper Technology 3. Lack of Funds
External factors	**Opportunity (O)** 1. Government support 2. New Technology strategic alliance	**Threats (T)** 1. Price Fluctuation 2. Market condition 3. Political meat

   A project which utilises the positive aspects like strength and opportunity is identified and a project with negative aspect like weakness & threats is dropped.

2. Brain Storming

   In this technique generally the top management and concerned individuals discuss together to reach generation of project ideas and Project Identification.

2. Pre-feasibility Studies

After the identification stage, the project-ideas are subjected to filtration through pre-feasibility study. In this stage more elaborate study is required than opportunity study stage. This is an intermediate stage between an opportunity study & feasibility study.

- This stage is recommended when project formulation & techno-economic feasibility study are costly and time consuming.
- This stage usually should not take more than 3 months. On the bases of this study investor should be able to decide.
  (a) Whether project can be accepted or rejected.
  (b) Project requires a detailed analysis i.e. feasibility study or
  (c) Some aspects of the project need to be specially investigated.

Outline of Prefeasibility Study for Industrial Projects:

(i) Executive summary-brief description and objectives of the proposal

(ii) Project background and history

(iii) Analysis of demand and supply

(iv) Analysis of input-requirement of raw material, tools and spares

(v) Location and site

(vi) Engineering and technology

(vii) Organisation

(viii) Manpower Study

(ix) Execution schedule and methodology

(x) Financial and economic evaluation

(xi) Assessment of environmental impact

3. Feasibility Studies

- Feasibility study is an analysis of how successfully a project can be completed. In feasibility analysis we study
    - Technical feasibility
    - market demand
    - financial feasibility
    - economic feasibility
    - social feasibility and environmental impact assessment.
- As compared to prefeasibility study, the feasibility study involves much more rigorous study and involves specialized skills of a higher order for techno-economic criteria.
- Even though the basic frame work is similar but it is based on more reliable data gathered through survey, research and pilot plant study.
- Even if a feasibility study report has covered all aspects, it may still lead to an incorrect project decision if
    (a) Project formulation is based on inadequate data.
    (b) The data is unrealistic,
    (c) The report has been deliberately tailored to fail acceptance criteria.
    (d) Unknown or unpredictable factors occur.

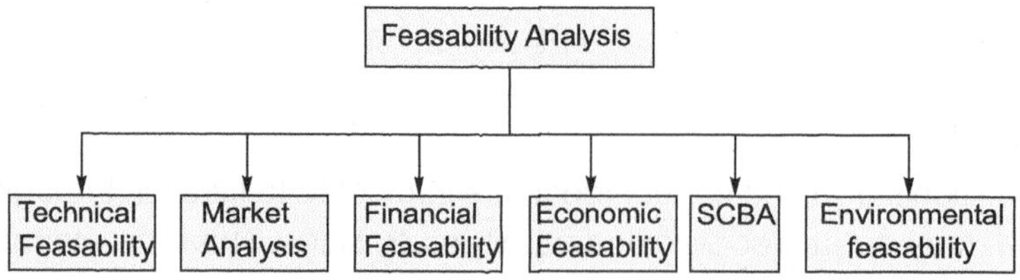

3.2 PROJECT APPRAISALS

3.2.1 Technical Analysis/Technical Appraisal

This broadly studies the technical details of the project under consideration such as process, technology-scale of operation, raw material etc. It is distinct from commercial, financial and managerial feasibility and it may vary project to project. The common aspects to be analysed are:

1. Objectives
2. Location and site
   - Identify as many as possible locations initially and choose the one which meets most of the requirements.
   - In some of the projects like resource oriented projects involve geological analysis also (Ex: Mining and Mineral projects)
   - The location decision should be made after giving due consideration to benefits offered by Govt. and local bodies such as tax concessions, capital loans, subsidies etc. For this purpose location cost index can be used for different cities.

   Location Cost Index (LCI)

   $$\text{LCI for location A is} = \frac{C_A}{C_B} \times 100$$

   where C_A = cost of setting of plant at location 'A'

   C_B = cost of setting of plant at location 'B'

   If LCI < 100, then project location should be preferred.
3. Plant size
   - Plant size should be optimum. The size of plant may have impact on break even, loss and profit.
4. Technology
   - Same services/products can be obtained by using different technologies, for example solar energy or Hydel or thermal energy, manual or mechanised services, etc.

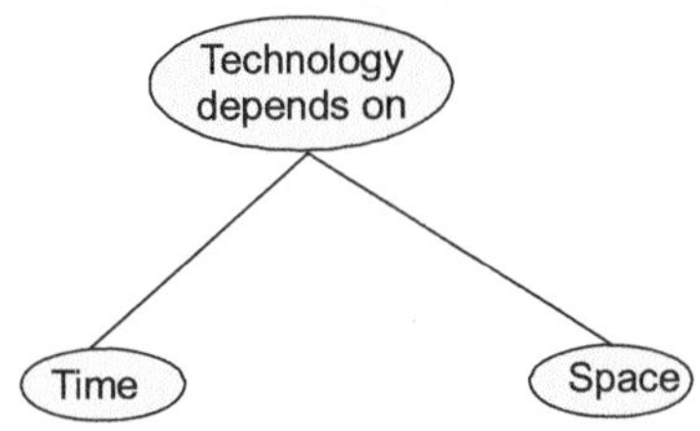

   - A technology is considered appropriate if it is assessed satisfactorily under specific situation of a project.
   - Technology is Time & Space dependent i.e. at time increases, technology may become inappropriate & as location changes technology may become inappropriate.

The important aspects to be considered in selection of technology are:

- Specifications of the task
- Ease or availability
- Cost involved
- Operation and maintenance
- Gestation period for development
- Adoptability or simplicity
- Safety
- Environmental and Social aspects
- Time horizon and future expandability

5. Design, Layout and Machinery
   - The design should be such that it is viable operating entity which not only works but works harmoniously under minimum cost and sustainable environment.
   - The selection of machinery and plant layout should meet industrial and safety standards.

6. Construction process

   This should be considered in feasibility study in terms of five aspects.

   (a) Methodology to be followed i.e. whether it is capital intensive or otherwise.

   (b) It is to be done in house or out-sourced.

   (c) Procurement method (purchase of material/machinery)

   (d) Sequence and scheduling of time to form bar chart/PERT network etc.

   (e) Assessment of financial implications.

7. Operation inputs

   These are related to operation phase but need to be identified during feasibility stage to examine technical feasibility

   The main inputs to be examined are:

   - Raw material
   - Processed material
   - Components and sub-assemblies
   - Spare parts
   - Water, gas and fuel supply

   Questions to be considered in this stage?

   - Is it proven/state-of-the-art/obsolete technology?
   - Whether the technology know-how is locally available or likely to be imported?
   - Do you visualize any problem with regard to scaling up operations?

Break Even Analysis

- Break even analysis starts with dividing the costs into two broad heads, *viz.,* fixed costs and variable costs.

- **Fixed Costs:** All projects incur certain costs that are fixed in nature. These costs remain constant irrespective of the changes in the volume of output. Following are some of the fixed costs:

  Rent payable for land,

  Factory/office premises

  Insurance premium on fixed assets

  Interest payable on long-term borrowing/deposits

  Administrative expenses

  Annual Maintenance Charges payable for machinery maintenance

  Depreciation

  Property tax etc.

- **Variable Costs:** These are those costs that vary directly with the level of output. Raw material cost is a variable cost since it depends on the level of output. Some other variable costs are as under.

  Consumable stores

  Power, fuel, water charges

  Selling expenses etc.

  At the Break even point, there is no profit and no loss, thus

  Sales realization (Revenue) = Fixed cost + Variable cost

  θ = Angle of incidence. Greater θ means greater rate of profit.

- **Contribution:** Contribution is the difference between the sales realisation and the variable cost. Excess of selling price over variable cost is called by the term 'contribution' since it contributes towards fixed cost and profit.

  For example, if the selling price per unit of a product is Rs. 12 and the variable cost per unit is Rs. 7, the contribution per unit is Rs. 5 (12-7) *i.e.,* one unit of production contributes Rs. 5 towards fixed cost and profit. If the fixed cost is say, Rs. 50.000. The BEP (in terms of no. of units) is 10,000 [50,000/5]. This means that at the level of output of 10,000 units. The contribution covers exactly the fixed cost. Hence, this is a no-profit, no-loss situation. Contribution first goes to meet the fixed cost up to BFP and subsequent to reaching of BEP, contribution adds to profit.

$$\text{BEP (in terms of no. of units)} = \frac{\text{Fixed price}}{(\text{selling price per unit} - \text{Variable cost per unit})}$$

$$= \frac{\text{Fixed price}}{\text{Contribution per unit}}$$

$$\text{B.E.P. (in terms of rupees)} = \frac{\text{Fixed cost}}{\text{Contribution per unit}} \times \text{Selling price per unit}$$

EXAMPLE 3.1

The details of production costs and revenues of a project are as under:

Total cost	**Rs. 65,000**
Fixed cost	**Rs. 25,000**
Sales (8,000 units)	**Rs. 80,000**

Find the BEP in terms of no. of units. What should be the output if the profit desired is Rs. 20.000?

Solution:

Variable cost (for 8000 units) : Total cost – Fixed cost

: Rs. 65,000 - Rs. 25.000

: Rs. 40,000

Variable cost per unit : Rs. 40,000/8,000

: Rs. 5

$$\text{BEP (in terms of no. of units)} : \frac{\text{Fixed cost}}{(\text{Selling price per unit} - \text{Variable cost per unit})}$$

$$: \frac{25000}{10-5}$$

: **5000 units**

Output for a desired profit of Rs. 20,000

Let 'X' be the no. of units required to be produced for achieving a profit of Rs. 20,000

$$\begin{aligned} \text{Profit} &= \text{Sales} - \text{Total cost} \\ &= \text{Sales} - (\text{Fixed cost} + \text{Variable cost}) \\ &= (10X) - (25{,}000 + 5X) \\ \text{i.e.,}\quad 20{,}000 &= 5X - 25{,}000 \\ 5X &= 45{,}000 \\ X &= 9{,}000 \end{aligned}$$

Therefore, the output required for achieving a profit of Rs. 20,000 is **9,000 units.**

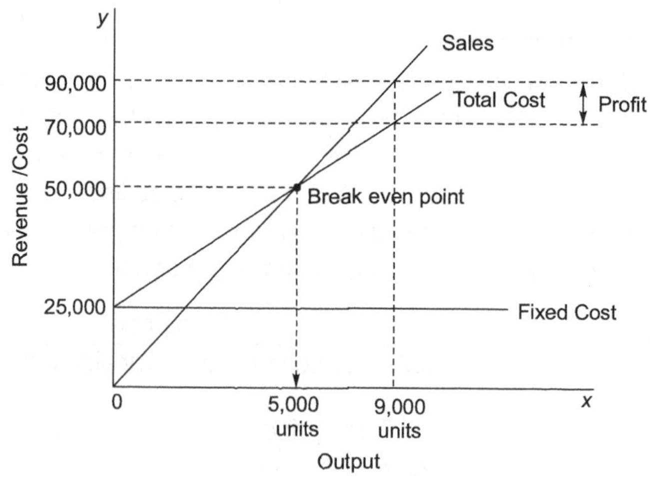

3.2.2 Market and Demand Analysis

It is to establish and understand the nature of the need. It is called demand forecasting. The DPR should lay down following:

(a) Demand forecasting
(b) Economic indicators
(c) Comparative Prediction
(d) Nature of product and services
(e) Secondary data sources
(f) Primary data sources
(g) Estimation of market share
(h) Marketing strategy
(i) The product - Market Posture
(j) Product, Price, Promotion and Place (4Ps)
(k) Product launch strategy
(l) Marketing organisation
(m) Marketing schedule
(n) Other marketing options

Market and Demand Analysis are required to meet planning and decision making. Following are specifications needed for planning:

(i) For short term planning decisions (Less than one year), the data required is specific items and their demand.

(ii) For Medium term planning (One to three years), the data required is aggregate demand of technology and site selection.

(iii) For long term planning (Four years or more), broad data and technology is required.

Demand Forecasting:

Time of demand forecasts varies from 1 to 5 years and demand forecasting can be made at the following levels:

1. **Firm level:** It is micro-level forecasting based on specific frim and its products.
2. **Industry level:** It is based on industry as a whole for the region. These forecasts may be taken by group of companies or trade associations.
3. **National level:** It includes parameters like national income, index of industrial/agricultural production. It is required for governmental decisions.
4. **International level:** It is required for multinational companies.

Criteria for a Good Forecasting:

A good forecasting should have following attributes:

(i) Accuracy

(ii) Plausibility (reasonable and consistent)

(iii) Economical

(iv) Less time taken

(v) User friendly and availability

(vi) Durability

(vii) Flexibility

Steps Used in Demand Forecast:

1. Determine the use of forecast
2. Select the item to be forecast
3. Determine the time horizon of the forecast
4. Select forecasting method
5. Gather data
6. Make forecast
7. Validate and implement results

Do you know?: Forecasting is an estimate of future events and trends on the basis of past data and projected for future. Whereas **Prediction** is an estimate of future events and trends in subjective manner without taking any past data.

Methods of Demand Forecasting

To facilitate proper and reliable appraisal of investment proposal, we require a reasonably accurate forecast of demand. Following methods are used for compiling and computing demand forecasts.

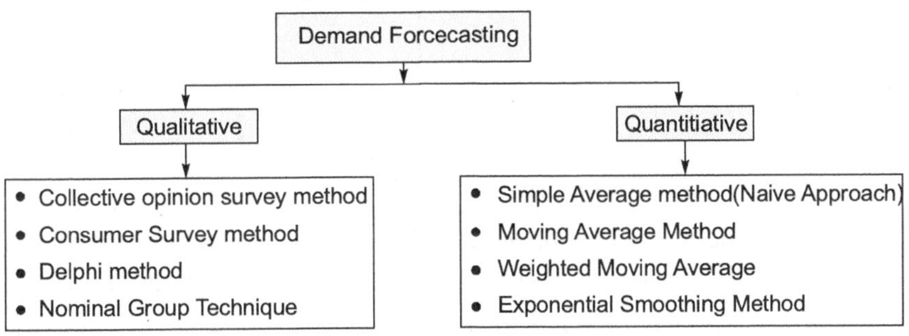

(a) Qualitative Methods

1. Collective Opinion Survey Method (Sales force composite method)

It is based on market survey i.e., opinion based on people of marketing and sales.

This method forms the basis of market analysis and demand forecasting. Although it is simple, direct, first hand and most acceptable, but it suffers from following weaknesses.

(a) Estimates are based on personal judgement hence may be biased.

(b) It is difficult at national level.

(c) There may be error or lack of accuracy.

(d) There may be lack of experience.

Note : This method may be used for long term forecast and for new products or new variants of existing products.

2. Survey of Customer intention method/Consumer Survey method

It is based on ***"what customer intend or prefers to buy"***. It can be done by personal contact or printed questionnaire or mail.

This method is preferred when bulk mass is required from institutions and industrial buyers and only few of them have to be contacted, because the demand is similar type.

3. Delphi Method

It is group process, based on panel of experts which may include ordinary employees or industry experts, it aims at achieving a 'consensus' of the members.

It is based on:

- Analysing economic conditions
- Carrying out sample surveys
- Conducting opinion polls

Based on above, demand forecast is done as follows:

Step-1: Coordinator sends out a set of questions in writing to all the experts co-opted on the panel who are requested to write back a brief prediction.

Step-2: Written predictions of experts are collected, edited and summarized together by the Coordinator.

Step-3: Based on the summary, Coordinator designs a new set of questions and gives them to the same experts who answer back again in writing.

Step-4: Coordinator repeats the process of collecting, editing and summarizing the responses.

Step-5: Steps 3 and 4 are repeated by the Coordinator to experts with diverse backgrounds till consensus is reached.

Note: In Delphi techniques the members identity is generally not disclosed so as not to have any halo effect and collect unbiased responses.

4. Nominal Group Technique/Jury of Executive Opinion

It is an improvement over Delphi method. A panel of 7 to 10 experts is formed to interact and discuss all the suggestions in descending order as per following procedure.

(i) Experts sit around a table in full view of one another and are asked to speak to each other.

(ii) Facilitator hands over copies of questionnaire needing a forecast and each expert is expected to write down a list of ideas about the questions.

(iii) After everyone has written down their ideas, facilitator asks each expert to share one idea from his own list with the group. The idea shared is written on the 'flip chart' which everyone can see.

(iv) Experts give ideas in rotation until all of them are written on the 'flip chart'. No discussion takes place in this phase and usually 15 to 25 ideas emerge from this format.

(v) In the next phase, experts discuss ideas presented by them. Facilitator ensures that all ideas have been adequately discussed. During discussions similar ideas are combined and paraphrased appropriately. This reduces the number of ideas.

(vi) After completing group discussions, experts are asked to give in writing ranks to ideas according to their perception of priority.

(b) Quantitative Methods

There are two models in quantitative forecasting:

(a) Time series model

(b) Casual model

Time series is a evenly spaced numerical data approach which is obtained by observing response at regular interval. Past collected data are used for future forecast.

Where as in casual model, mathematical technique known as regression analysis is used that relates dependent variables (ex. demand) to independent variables (ex. price and advertisement) through linear equations.

1. **Simple Average Method / Naive approach**

   Simple average of all past data is figured out for every month of the last 12 months.

$$\text{Simple average} = \frac{\text{Sum of Demands of all periods}}{\text{Number of periods}}$$

2. **Moving Average Method**

   It is better than simple average method because the obsolete and old data is discarded where as the trend for most recent period is used.

$$\text{Moving average} = \frac{\text{Sum of Demand of chosen period}}{\text{Number of chosen period}}$$

3. **Weighted Moving Average Method**

   It is further improved in which weighted moving average allows to varying weightage for demand in old periods.

   Weighted Moving Average = $W_1 \cdot D_1 + W_2 \cdot D_2 + \ldots\ldots W_n \cdot D_n$

   where $W_1, W_2 \ldots W_n$ are the weightage of the different periods in fraction such that

   $W_1 + W_2 + \ldots. + W_n = 1$

   $D_1, D_2 \ldots. D_n$ are the actual demand of various periods.

NOTE : This method has the advantage that it allows to compensate for known trends or seasonality can be fitted by appropriate coefficients.

4. **Exponential Smoothing Method**

   It has become popular due to software invention. It is distinguished from weighted average method by the way of assigned weightage to each past period. The pattern of weightage is exponential.

$$\textbf{Forecast of next period} = \begin{matrix} w \times \text{Actual demand of} \\ \text{most recent period} \end{matrix} + \begin{matrix} (1 - w) \times \text{Demand forecast of} \\ \text{most recent period} \end{matrix}$$

$$Ft = w(Dt - 1) + (1 - w)(Ft - 1) \quad \ldots(i)$$

where '*t*' is the time period and value of '*w*' varies between 0 to 1.

Continuing further, we have

$$Ft - 1 = w \times (Dt - 2) + (1 - w)\,(Ft - 2) \text{ and so on}$$

Substituting these values of $Ft - 1$ into equation (i) above

$$Ft = (Dt + 1) + w(1 - w)\,(Dt - 2) + w\,(1 - w)\,(Ft - 2)$$

This can be rewritten as

$$Ft = w(1 - w)\,(Dt - 1) + w\,(1 - w)\,(Dt - 2) + w(1 - w)\,(Dt - 3)$$

Successive weightages are w, $(1 - w)$, $w\,(1 - w)$, etc.

For $w = 0.3$, these successive coefficients will be 0.3, 0.21, 0.147 etc.

When forecast have to be worked out successively for different periods, it is a simple computation as below:

S.No.	Month	Forecast	Demand	Weighted
1.	January	200	300	w = 0.7
2.	February	0.7 × 300 + 0.3 × 200 = 270	350	w = 0.7
3.	March	0.7 × 350 + 0.3 × 270 = 326	400	w = 0.7
4.	April	0.7 × 400 + 0.3 × 326 = 378	?	

Choice of coefficient of smoothing is critical:

(i) a value of 0.7 to 0.9 may be more appropriate for new products or items for which demand is shifting significantly.

(ii) a value of 0.1 to 0.3 denotes stable trend in demand and when demand is somewhat unstable.

(iii) value of 0.4 to 0.6 may be used for obtaining more accurate forecasts.

These methods are simple and with the speed and accuracy of computers, are being increasingly used for determining 'quick and easy' forecasts within reasonable budgets.

NOTE: Method number 5 to 8 are statistical methods.

(c) Some Other Methods

1. Adoptive Exponential Smoothing
2. Exponential Smoothing with trend and seasonal data
3. Double exponential smoothing
4. Projection by curve fitting
5. Regression Analysis
6. Economic models
7. End Use Method
8. Leading Indicator Method
9. Chain-Ratio Method

Errors in Demand Forecasting

- Forecasting error may be defined as the numeric difference between forecasted demand and actual demand.
- Any errors that are present in demand forecast will impact on reliability of the economic appraisal.
- There can be three types of errors in demand forecast
  1. Measurement and data error
  2. Model specification error
  3. External error

Following aspects are important in measuring the effectiveness of any method of forecasting:

(a) Mean Absolute Deviation

- It is measure of the forecast error irrespective of to the direction of the error, i.e., negative or positive. It is given by the expression:

$$\text{Mean absolute deviation} = \frac{\text{Sum of absolute values of forecast errors for all periods}}{\text{Number of periods}}$$

- Values are taken without algebraic signs. Mean Absolute Deviation (MAD) is the average of the Absolute Values of forecast Errors. Forecasts errors are, by themselves, normally distributed and there is a relationship between MAD and Standard Deviation. If the forecast is reasonably accurate, forecast errors will be quite smooth. The above measure is then referred to as Smoothed Mean Absolute Deviation (SMAD). In such events,

$$\text{Standard Deviation} = 1.25\ \text{SMAD}$$

(b) Bias

- Bias is less commonly used measure of forecast errors. It takes into account the direction of errors and sum of errors is the algebraic sum of deviating and is expressed as:

$$\text{Bias} = \frac{\text{Algebraic sum of forecast errors of all periods}}{\text{Number of periods}}$$

- If forecasts are repeatedly overestimates, bias will have a positive value. Similarly, if forecasts are consistently underestimates, Bias will have a negative value.
- For a very accurate forecast, both MAD and bias will be zero. It is preferable to chose a method of forecasting wherein MAD can be controlled and brought nearer to zero.
- Lowering MAD will reduce Bias.

Forecast Errors

Forecast Accuracy Measure		Formula		Description
Mean Square Error	MSE	$MSE = \frac{\Sigma(F_t - Y_t)^2}{n}$	(4)	Forasmuch any error is being raised with teh square. So this way highlights the significant error values. This feature is quite significant because forecasting methods with approximations of bias are frequently more suitable than the method which gives not only negligible errors but significant.
Mean Absolute Deviation	MAD	$MAD = \frac{\Sigma F_t - Y_t}{n}$	(5)	It is similar to standard deviation but the formula of estimation is less difficult to apply for time series. The usage is advisable when the forecast bias must be estimated using the same evaluation units as forecast factor is evaluated.
Mean Forecast Error	MFE	$MFE = \Sigma(F_t - Y_t)$	(6)	Very often it is very important to estimate whether the forecast method has a systematic error i.e. the present forecast value is always major (or minor) than time series value. In this case the mean forecast error is being used. If the systematic bias does not exist the MFE value will be equal to zero. If the forecast value is signally negative the forecast method overestimates trend series. If the systematic bias is signally positive the forecast method generates major values than time series.
Mean Absolute Percentage Error	MAPE	$MAPE = \frac{1}{n}\Sigma\left\|\frac{F_t - Y_t}{Y_t}\right\|$	(7)	The Mean Absolute Percentage Error is useful when assessing the forecast error an important factor is the estimated value. MAPE estimates the size of bias comparing with time series values. This fact is very important when the times series value is quite large.
Tracking Signal	TS	$TS = \frac{\Sigma(F_t - Y_t)}{MAD}$	(8)	Tracking signal is the method to control the forecast accuracy. New data is compared with forecast time series and adequacy is evaluated.

(c) Costs of Forecast Errors

- Important decisions (short-term, medium and long-term) are based on forecasts. Large errors of forecasts can lead to costly mistakes, particularly, when used for evaluating investment proposals which are long-term and nearly irreversible.

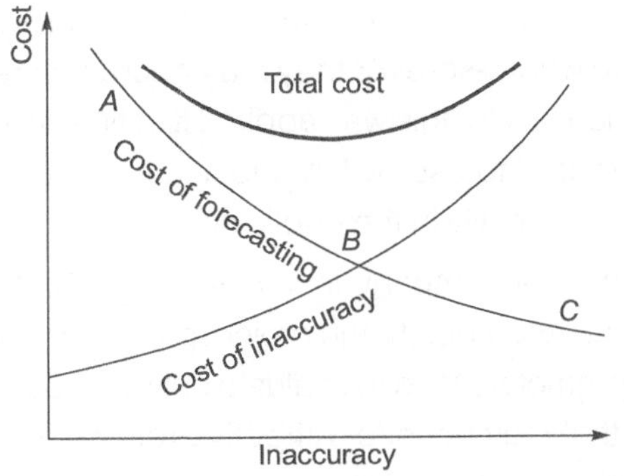

A – Regression and other causal methods
B – Sophisticated statical method
C – Simple statical method

Fig. *Costs and accuracy of forecasts*

- Error may be either in direction or in magnitude. In some situations, errors in direction can be critical while in other cases, error in magnitude may be more damaging.

(d) Costs and Accuracy of Forecasts

- It is possible to establish relationship between percentage error of forecast and the costs involved in selecting, adopting, operating and maintaining that particular method of forecasting.
- If we use qualitative methods of forecasting, these would entail less effort and expense but would also yield forecasts with higher percentage error.

  If we use time-series and other mathematical techniques, costs would increase but it would also reduce the percentage of errors.
- Using more sophisticated statistical tools shall further improve the accuracy of forecasts. Correspondingly, cost of operating and maintaining these forecasting techniques will vary as shown in the graphologists would be low when errors are low but costs would increase if we have to base our planning and decision making on account of larger margin of errors.
- Optimal region, corresponding to minimum overall costs, would be somewhere in the middle of the Total Cost Curve–usually nestling between simple and sophisticated statistical techniques which can be compiled on a computer.

3.3 FINANCIAL ANALYSIS /FINANCIAL FEASIBILITY

Financial analysis refers to an activity of assessing financial statements to judge the financial performance of a company. It helps in assessing profitability, solvency, liquidity and stability. It has been separately dealt in next chapter.

3.4 ECONOMIC ANALYSIS (ECONOMIC FEASIBILITY)

Economic analysis is done from the viewpoint of society or economy as a whole. The evaluation is done from a wider angle not merely in financial terms. Economic analysis of projects should cover whether.

- It fits into national priorities
- It contributes to the development of that sector of economy
- Benefits justify the consumption of scarce resources of the nation
- The project has to be evaluated for its economic viability

3.5 SOCIAL COST BENEFIT ANALYSIS (SCBA)

This concept has evolved over years since 1844 when Jules Dupuit, A French engineer, referred to it in his paper on the measurement of "Utilities of Public Works'. In 1936, Flood Control Act of USA provided that a project should be deemed feasible only if **sum-total of benefits to whom-so-ever they may accrue exceed the estimated costs,** highlighting the social nature of investment decision. In UK, this was applied first time in 1917 for evaluating Motorway project, and nationalized industries were directed to use SCBA. India is a pioneer in the third world countries and has applied SCBA in appraising projects especially in the public sectors.

- The main objective of an individual, a firm or a company in investing on a project is to earn the maximum possible returns for the investment. Accordingly, the project promoters are solely interested in wealth maximization. Hence the project promoters tend to evaluate only the commercial profitability of a project. But there are some projects that may not offer attractive returns as far as commercial profitability is concerned, but still such projects are undertaken since they have social implications. Such projects are public projects like road, railway, bridge and other transport projects, irrigation projects, power projects etc., for which socio-economic considerations play a significant part rather than mere commercial profitability. Such projects are analysed for their net socio-economic benefits and the profitability analysis of such projects is known as national profitability analysis, which is nothing but the socio-economic cost-benefit analysis done at the national level.

- Every project imposes certain costs to the nation and produces certain benefits to the nation. The costs may be of two types *viz.,* direct costs and indirect costs. So also, the benefit derived from any project will also be of two types, *viz.,* direct benefits and indirect benefits.
    (a) The social cost-benefit analysis takes into account the real cost of inputs i.e., the cost of inputs had they not been subsidised. Accordingly, the required adjustments to direct costs of inputs are made for social cost-benefit analysis.
    (b) Whereas in the direct cost and direct benefits are considered for commercial profitability analysis, socio-economic cost-benefit analysis takes into account the indirect costs and indirect benefits to the nation. While a nation bears the indirect cost, the people of the nation enjoy the indirect benefits.

Objectives of Social Cost Benefit Analysis (SCBA)

SCBA aims to appraise the total impact that a project will have on an economy. Accordingly, SCBA focuses on the following objectives that a project is expected to fulfill.

- Contribution of the project to the GDP (Gross Domestic Product) of the economy.
- Contribution of the project to improve the benefits to the poorer sections of the society and to reduce the regional imbalances in growth and development.
- Justification of the use of scarce resources of the economy by the project.
- Contribution of the project in protecting/improving the environmental conditions.

Methods or Approaches to Social Cost Benefit Analysis (SCBA)

There are two approaches:

(a) UNIDO Method (United Nation Industrial Development Organisation)

UNIDO gave SCBA for projects in developing countries in 1960s & followed it by a detailed guideline in 1978, which suggests five stages as below:

- UNIDO method measures cost benefit in terms of domestic prices
- Calculation of financial profitability of the project measured at market prices
- Calculations of net benefits of projects in terms of economic prices or shadow prices.
- Adjustment for the impact of project on savings and investments.
- Adjustment for the impact of project on income distribution.
- Adjustment for the impact on the merit goods and demerit goods for which social values differ from monetary values.

NOTE

What is Shadow Price?

If price does not reflect the actual value of a good or commodity or no market value for a good exit, then shadow pricing can be used. Shadow price is a proxy value of a good, often defined by what an individual should pay to buy one unit of good. It may be greater, equal or less than market price.

Ex. If there is unemployment then shadow wage of a labour is less than market wage.

(b) Little—Mirrless Method (L-M Method)

- L-M's approach measures costs and benefits in terms of international price (also represented as border price) as against UNIDO method that measures costs and benefits in terms of domestic prices. The argument in favour of using international price as shadow price is that it helps offset the fluctuations in domestic price and justifies the project from the economy's point of view. It also helps to improve productive efficiency and trade efficiency.
- **L-M shadow price for traded goods:** For goods that are exported, the shadow price is the F.O.B. (Freight on Board) price while for goods that are imported, the shadow price is the C.I.F. (Cost, Insurance and Freight) price.
- **L-M shadow price for non-tradeable goods:** Non-tradeable (excluding labour) include goods like land, building and services like power, internal transport etc., Shadow price for non-tradeables (except labour) is arrived in terms of marginal social cost and marginal social benefit. The marginal social cost and marginal social benefit are estimated in terms of border prices. By using border prices even for non-tradeables, L-M approach only ensures that a common yardstick is used for valuing both tradeables and non-tradeables.
- It is infect difficult to measure the marginal social cost and the marginal social benefit of non-tradeables and that too at the border prices. L-M approach suggests breaking down the cost component of non-tradeable into tradeable component, labour and residual component and applying a suitable Standard Conversion Factor (SCF) for tradeable and residual components and using Standard Wage Rate (SWR) for the labour component. The difficulty however is the selection of an appropriate SCF since this involves an element of subjectivity.

Social Cost Benefit Analysis in India

India has been one of the pioneers in the use of SCBA for public sector projects which were, at one time, at the commanding heights of the economy. ICICI was perhaps the first financial institution to introduce economic appraisal of projects. IFCI and IDBI followed with similar approach. The method followed by them, is a simplified version of L-M method. Project Appraisal Division (PAD) of Former Planning Commission (Now NITI Aayog), Government of India, also followed economic analysis, they call it social **profitability analysis,** of all industrial projects. They also follow a simplified version of L-M method. The key elements of analysis are:

- Tradable inputs are valued at international prices
- Taxes, duties etc., Transfer costs, are ignored
- Non-tradable items like power & equipment are valued on marginal cost
- Foreign exchange for inputs and outputs is valued at a specified premium
- Social wage rates are applied to unskilled and semiskilled labour.
- Unit of currency in savings is rupees rather than foreign exchange.

Unfortunately, inspite of pioneering work done in India, there have been serious gaps in arriving at investment decisions which are financially and economically sound owing to political expediencies.

Indian financial institutions use the following indicator for cost benefit analysis:

- Economic rate of return or internal rate of return
- Domestic resource cost
- Effective rate of protection

Economic Rate of Return (ERR):

- Economic Rate of Return measures the rate of return to the society. The current market prices and costs, when taken into account to arrive at the financial viability of a project does not represent the true value of the project from economic point of view.

- This method of financial evaluation may be of help only for private project promoters and not for social projects.
- ERR is the rate of discount that equates the economic cost of the project to the economic benefits that would occur during the life of the project. For the calculation of ERR, shadow prices are used both for inputs (costs) and for outputs (benefits) instead of market prices. Social projects are compared on the basis of ERR before choosing one for implementation. A project may have a better ERR, though it may have a comparatively lower financial rate of return and vice versa.

Domestic Resource Cost (DRC):

- Domestic Resource Cost measures the domestic resource cost of manufacturing a product as against the cost of importing it.
- DRC attempts to measure the cost of import substitution. If the production of a product contemplated by the project results in import substitution, there is saving of foreign exchange. By measuring the DRC of such a project, an attempt is made to calculate the amount of domestic resources required to be expended for saving the foreign exchanges.
- Also, a commodity produced may be exported which results in foreign exchange earnings. Thus, when a product produced by a project substitutes import, foreign exchange is saved due to reduced imports and when a commodity produced is exported, foreign exchange is earned to the extent there is increase in exports. Setting up of the project and manufacturing the product may involve import of machinery and/or raw material which will result in outflow of foreign exchange and this effect is to be accounted for in the calculation of DRC.

$$\text{DRC} = \frac{\text{Value added at domestic prices}}{\text{Value added at world prices}} \times \text{Exchange rate}$$

- DRC may be taken as a measure of total rupees spent to save a unit of foreign currency. (This definition is applicable where the goods produced are import substitutes.) Similarly, DRC may also be taken as a measure of the total rupees spent to earn a unit of foreign currency. (This definition is applicable where goods produced are meant for exports).
- DRC acts as an indicator to decide if it is worthwhile to manufacture a product in the country. *If the DRC is less than the exchange rate of the unit of foreign currency, it indicates that manufacturing the product indigenously is advisable.* This is because, it is possible to manufacture the product at a cost which is lower than the cost of foreign exchange. Hence, it is worthwhile to implement the project.

EXAMPLE 3.2

Let,

Value added at domestic price per unit of the proposed product	:	Rs. 50
Value added at world price per unit of the proposed product	:	Rs. 100
Exchange rate	:	1 US$ = Rs. 50
Therefore, Domestic Resource Cost	:	(50/100) × 50
	:	Rs. 25

Since the DRC (Rs. 25) is less than the exchange rate (Rs. 50), it is worthwhile to implement the project for the manufacture of the proposed product.

NOTE

It is not worthwhile to manufacture a product, if the DRC is more than the exchange rate. However, keeping in view the scarcity of foreign exchange and social/economic benefits like the creation of employment opportunity, utilization of indigenously availably resources etc., the Government of a country may decide to set up public sector projects even if the DRC is slightly higher than the exchange rate.

Effective Rate of Protection (ERP):

- It measures the net protection offered to domestic projects. The domestic price of a product is affected by the concessions, tariffs, subsidies, quantitative controls etc., that are in force in a country. Thus, the costs of inputs that go into a product are affected by such controls and concessions. The competitive strength of a product in the world market is revealed only by costs and prices after excluding concessions and controls. In the absence of concessions and controls, the value added to a product when measured at domestic prices and when measured at international prices will be the same. If there is any difference in value addition when measured in terms of domestic prices and international prices, it indicates the measure of protection.

$$\text{ERP} = \frac{\text{Value added at domestic prices} - \text{Value added at world prices (Free trade)}}{\text{Value added at world prices}}$$

- ERP of zero indicates that the project does not enjoy any protection from international competition.
- Positive value of ERP indicates that the project enjoys protection from international competition. Higher the positive value of ERP, higher the protection enjoyed by the product.

EXAMPLE 3.3

	Domestic prices (Rs. in crores)	World prices (Rs. in crores)
Cost of inputs	550	570
Sales realization	620	610
Value added	70	40

Note: *Value added = Selling price – Cost of inputs (intermediate goods)*

$$\text{Effective Rate of Protection} = \frac{70-40}{40} = 0.75 = 75\%$$

The protection available for the product given in the example is 75%. In case the ERP is equal to 1.00, the protection available is 100%.

- Projects with a higher ERP require careful analysis as to whether the protection available against international competition is likely to continue. It may be possible that the project may have a higher ERP because of subsidized cost of inputs. If in future, there is possibility of the Government withdrawing the subsidies presently allowed on inputs, the ERP will come down and the project may get exposed to international competition.

3.6 ENVIRONMENTAL ANALYSIS/LOGICAL FEASIBILITY

Here the study of impact of the project on the environment is done to ensure the pollution that is sound, air, water etc. and deforestation due to the project do not impact the environment adversely.

The environmental feasibility study considers both human and environmental health factors.

Social Impacts on communities affected by the project include, for example, requirements for resettlement and the associated impact on quality of life and livelihoods, and impacts related to environmental alteration (e.g. on health and livelihoods).

Environmental impacts on the project location and in associated areas (e.g. downstream, ground water or ambient air) include effects on environmental resources due to alterations or pollutants.

It will often by a mandatory regulatory requirement for assessments of social and environmental impacts to be carried out during infrastructure project development. The scope of social and environmental studies can cover:

- Quantifiable social and environmental costs and benefits.
- Non quantifiable social and environmental costs and benefits
- Options for mitigating adverse impacts and the cost of mitigation.

The secondary effects should be included in the assessment. Public consultation is often a part of the social and environmental feasibility process.

The analysis should identify what type of social and environmental impact studies are needed, and the type of permits and licenses required, and should take into account health and safety standards. This information will assist the sponsor with the preparation of tender documents if the project is taken to market, and will assist bidders with the preparation of risk minimizing bids.

The final assessment of environmental and social costs and benefits is an input to the economic assessment of the project. Therefore, in addition to being a requirement from a legal and regulatory perspective, the social and environmental analysis is an important part of the assessment of project's overall welfare impact, as captured in the economic analysis. Environmental and social analysis will also help in selection of appropriate technology for minimisation of risk due to adverse social or ecological impact of the project.

3.7 PROJECT APPRAISAL AND PROJECT SELECTION

It is essential to evaluate the proposed solution against its ability to solve the identified problem or need.The financial, the cost-effectiveness and the feasibility analyses will serve as the methods of project appraisal to approve the project.

Project Appraisal is a consistent process of reviewing a given project and evaluating its content to approve or reject this project, through analyzing the problem or need to be addressed by the project, generating solution options (alternatives) for solving the problem, selecting the most feasible option, conducting a feasibility analysis of that option, creating the solution statement, and identifying all people and organizations concerned with or affected by the project and its expected outcomes. It is an attempt to justify the project through analysis, which is a way to determine project feasibility and cost-effectiveness.

Types of project appraisal: The most common method of appraisal are financial and economic appraisal. In case of financial project appraisal, the company reviews the cost of the project and the expected revenues that will be generated by the project. This type of appraisal helps the company to prevent overspending on a project. It also helps in finding certain areas where alterations can be done for generating higher revenues. Under economic appraisal, the company mainly focuses on the total benefit of the project and less on the costs spent on the project. Other than these two types of appraisal, there are also other types of project appraisal which include technical appraisal, management or organizational appraisal and marketing and commercial appraisal.

Project Selection

Project Selection is a process to assess each project idea and select the project with the highest priority.

Projects are still just suggestions at the formulation stage, so the selection is often made based on only brief descriptions of the project. As some projects will only be ideas, you may need to write a brief description of each project before conducting the selection process.

Selection of projects is based on benefits of the project and feasibility of the project.

Benefits: A measure of the positive outcomes of the project. These are often described as "the reasons why you are undertaking the project". Fulfilling commitments made as part of national, regional or international plans and agreements.

Feasibility: A measure of the likelihood of the project being a success, i.e. achieving its objectives. Projects vary greatly in complexity and risk. By considering feasibility when selecting projects it means the easiest projects with the greatest benefits are given priority.

Project Selection Models

Non Numeric:

Sacred Cow - project is suggested by a senior and powerful official in the organization

Operating Necessity - the project is required to keep the system running

Competitive Necessity - project is necessary to sustain a competitive position

Product Line Extension - projects are judged on how they fit with current product line, fill a gap, strengthen a weak link, or extend the line in a new desirable way.

Comparative Benefit Model - several projects are considered and the one with the most benefit to the firm is selected

Numeric:

Most common measures numeric methods are Net Present Value (NPV), Internal Rate of Return (IRR), Payback Period, Return on Investment (ROI), Discounted Cash Flow. The methods have been discussed in next chapter.

3.8 PROJECT CHARTER

It is first document of the project.

It answers questions like who, what, whom, how.

The Role of Project Charter

Following are the roles of a Project Charter:

- It formally declares the presence of the project.
- It documents the reasons for undertaking the project.
- It clearly points out the objectives, goals and the constraints faced by the project.
- It has well documented various problem and their solutions.
- It recognises main stakeholder and gives authority to the project manager.

Benefits of Project Charter

- It improves and paves way for good customer relationships.
- It clearly defines the role of the project manager and stakeholders.
- Project charter also works as a tool that improves project management processes.
- By having a project charter, project sponsorship can also be gained.
- Project Charter recognizes senior management roles.
- Allows progression, which is aimed at attaining industry best practices.

Elements in Project Charter

For an effective charter project, it needs to address these key elements :

- Identity of the project.
- Time : the start date and the deadline for the project.
- People (all stakeholders) involved in the project.
- Set targets and define goals.
- Detailed description of a problem or an opportunity.

- The benefits and result (in product and performances terms) expected from the project.
- The expected date that the objectives is to be achieved.
- Clearly defined roles and responsibilities of the participants involved.
- Requirement of resources that will be needed for the objectives to be achieved.
- Proper communication with detailed know how.

The project charter is not only a tool that is used for planning projects but also a communication mechanism that acts as a reference. A well-planned project with an effective communication plan will definitely bring in success for the project undertaken at hand.

This makes the project charter very important document for the success of the project which must be known properly to all the members in project.

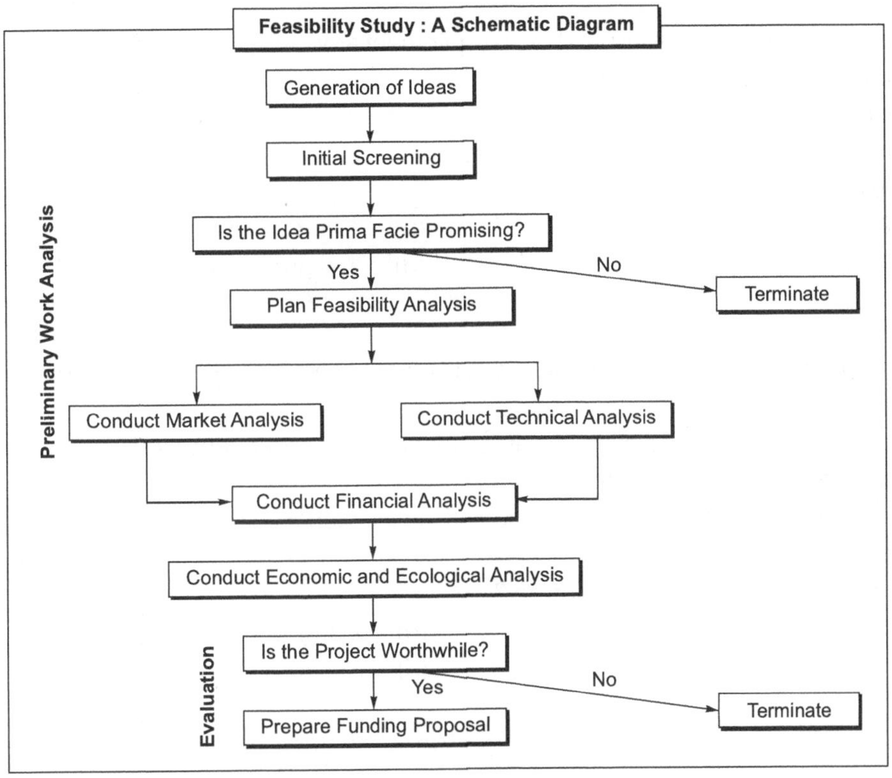

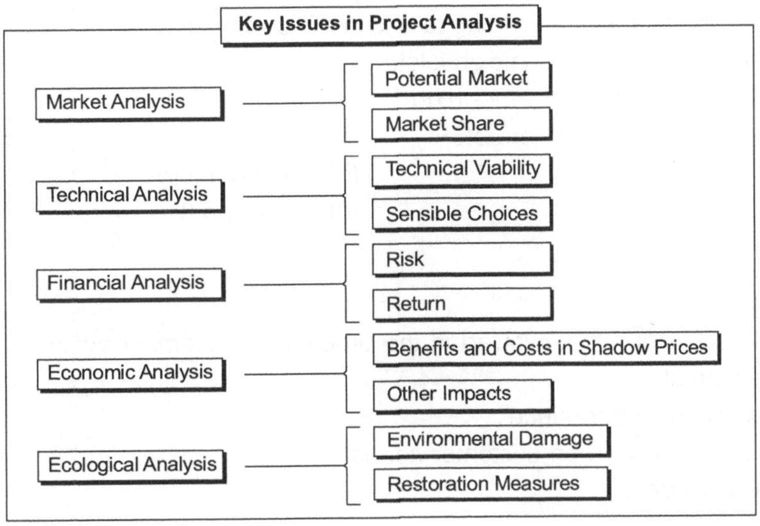

Previous ESE Prelims Questions

Q.1 Which of the following methods are considered as most favourable to production organizations liable for taxation?
1. Straight line method of depreciation.
2. Declining balance method of depreciation.
3. Sum of the years digits methods of depreciation.
4. Sinking fund method of depreciation.

Select the correct answer using the codes given below:
(a) 1 and 2 only (b) 2 and 3 only
(c) 3 and 4 only (d) 4 and 1 only [ESE-2017]

Ans. (b)

Q.2 Factors which govern the operating cost of an equipment are:
1. Purchase price of the equipment.
2. Depreciation due to regular use.
3. Cost of operation, maintenance and repairs.

Which of the above statements are correct?
(a) 1 and 2 only (b) 1 and 3 only
(c) 2 and 3 only (d) 1, 2 and 3 only [ESE-2017]

Ans. (c)

Q.3 Consider the following statements with reference to maintenance and service of the product after delivery to the customer:
1. Reliability analysis uses statistics of failures to estimate, understand and improve the performance of the equipment and its maintenance.
2. Reliability analysis records supports or guides the maintenance engineer in improving the performance of the product under service

Which of the above statements is/are correct?
(a) 1 only (b) 2 only
(c) Both 1 and 2 (d) Neither 1 nor 2

[ESE-2017]

Ans. (c)

Q.4 The 'Economic life' of a building is considered to be at an end
1. When the net income from the building fails to justify existence
2. When the building facilities become obsolete
3. When the capitalization rate becomes high

Which of the above statements are correct?
(a) 1 and 2 only (b) 1 and 3 only
(c) 2 and 3 only (d) 1, 2 and 3

[ESE-2017]

Ans. (a)

Q.5 Which of the following steps are involved in the product improvement cycle?
1. Sell it in the market
2. Determine quality of performance
3. Design the product based on customer needs
4. Test it in the laboratory

Select the correct answer using the codes given below:
(a) 1 and 3 only
(b) 2 and 4 only
(c) 1, 3 and 4 only
(d) 1, 2, 3 and 4 [ESE-2019]

Ans. (d)

Q.6 What is the break-even sale for the following products?

	Products		
	A	B	C
Sales (units)	5,000	6,000	4,000
Unit selling price (Rs.)	10	15	18
Unit variable price (Rs.)	6	4	13
Fixed cost of the product is (Rs. 20,000)			

(a) Rs. 90,000
(b) Rs. 80,000
(c) Rs. 60,000
(d) Rs. 40,000 [ESE-2019]

Ans. (d)

Q.7 Which one of the following is a viable alternative to term-loans and are instruments for raising debt finance by large publicly traded firms?
(a) Shares
(b) Debentures
(c) Asset loans
(d) Gold loans [ESE-2019]

Ans. (b)

Q.8 Phillip Kotlar argues that the 4 Ps which represent the seller's thinking more than buyer's thinking can be translated into the 4 Cs. Match the following:

4 Ps of Marketing Planning		**4 Cs of Marketing Planning**
A. Product	1.	Customer communication
B. Price	2.	Customer value
C. Place	3.	Customer costs
D. Promotion	4.	Customer convenience

Select the correct matching using the code given below:

	A	B	C	D
(a)	2	3	4	1
(b)	2	3	1	4
(c)	1	4	3	2
(d)	1	4	2	3

[ESE-2021]

Ans. (a)

THE FOUR P'S OF MARKETING:

Product: A product refers to an item that satisfies the consumer's needs or wants. Products may be tangible (goods) or intangible (services, ideas or experiences).

Price: Price refers to the amount a customer pays for a product. Price may also refer to the sacrifice consumers are prepared to make to acquire a product (e.g. time or effort), and includes considerations of customer perceived value.

Place: Place describes giving the customer access to your product, traditionally at a brick-and-mortar place or through the mail.

Promotion: Promotion refers to marketing communications (advertising, PR, or sales/promotions).

Q.9 Which one of the following policies is concerned with changing the supply of money stock and the rate of interest, for the purpose of stabilizing the economy at full potential output level?

(a) Commercial policy
(b) Fiscal policy
(c) Monetary policy
(d) Social policy

[ESE : 2022]

Ans. (c)

Commercial policy is an umbrella term describing the regulations and policies that dictate how companies in different countries can conduct commerce with each other. Commercial policy includes tariffs, import quotas, export constraints, and restrictions against foreign-owned companies operating domestically.

Fiscal policy refers to the use of government spending and tax policies to influence economic conditions, especially macroeconomic conditions, including aggregate demand for goods and services, employment, inflation, and economic growth.

Monetary policy is a set of tools that a nation's central bank has available to promote sustainable economic growth by controlling the overall supply of money that is available to the nation's banks, its consumers, and its businesses.

Fiscal policy is often contrasted with monetary policy, which is enacted by central bankers and not elected government officials.

Social policy is concerned with the ways societies across the world meet human needs for security, education, work, health and wellbeing. Social policy addresses how states and societies respond to global challenges of social, demographic and economic change, and of poverty, migration and globalisation.

Q.10 Consider the following statements :

The broad areas of corporate appraisal and the few important aspects to be considered under them are

1. Marketing and Distribution
2. Production and Operation
3. Research and Development
4. Project Rating

Which of the above statements are correct?

(a) 1, 2, 3
(b) 2, 3, 4
(c) 1, 2, 4
(d) 1, 3, 4

[ESE Prelims : 2024]

Objective Brain Teasers

Q.1 The reason that the schedule performance index (SPI) is shown as a ratio is to:

(a) Enable a detailed analysis of the schedule regardless of the value of the schedule variance
(b) Distinguish between critical path and noncritical path work packages
(c) Provide the ability to show performance for a specified time period for trend analysis
(d) Measure the actual time to complete the project

Q.2 If the cost variance is the same as the schedule variance and both numbers are greater than zero, then:

(a) The cost variance is due to the schedule variance
(b) The variance is favorable to the project
(c) The schedule variance can be easily corrected
(d) Labor rates have escalated since the project began

Q.3 Preliminary project report for a road project must contain

(a) the detailed estimated cost based on detailed design
(b) the several alternatives of the project that have been considered

(c) the soil survey, traffic survey, concept design and approximate cost
(d) the contract documents for inviting tenders

Q.4 Match **List-I** (Feasibility) with **List-II** (Emphasized on or calls for) and select the correct answer using the codes given below the lists:

List-I	List-II
A. Social	1. Optimizing marginal costs and benefits
B. Economic	2. Forward and backward linkages
C. Input	3. Interfaces
D. Co-ordination	4. Improving wealth

Codes:

	A	B	C	D
(a)	1	4	3	2
(b)	4	1	3	2
(c)	1	4	2	3
(d)	4	1	2	3

Q.5 Cost-benefit studies are essential to
(a) assess the total cost of the work
(b) ascertain the relevant escalation in prices
(c) monitor the expenditure
(d) evaluate the viability and worthwhileness of taking up the project

Q.6 All the following are project baselines that are generally part of the project management plan EXCEPT
(a) Technical (b) Scope
(c) Time (d) Cost

Q.7 The Delphi technique is a particularly useful method for identifying risks to:
(a) Present a sequence of decision choices graphically to decision makers
(b) Define the probability of occurrence of specific variables
(c) Reduce bias in the analysis and keep any one person from having undue influence on the outcome
(d) Help take into account the attitude of the decision maker toward risk

Q.8 Risk mitigation involves:
(a) Using performance and payment bonds
(b) Eliminating a specific threat by eliminating the cause
(c) Avoiding the schedule risk inherent in the project
(d) Reducing the probability and/or impact of an adverse risk event to an acceptable threshold

Q.9 What is the primary difference between a risk audit and a risk reassessment?
(a) A risk reassessment is conducted at the completion of a major phase; audits are conducted after the project is complete.
(b) Project stakeholders conduct risk audits; management conducts reassessments.
(c) Risk reassessments are regularly scheduled; risk audits are performed as defined in the project's risk management plan.
(d) There is no difference; they are virtually the same.

Q.10 Accurate and unbiased data are essential for perform qualitative risk analysis. Which one of the following should you use to examine the extent of understanding of project risk?
(a) Data quality assessment
(b) Project assumptions testing
(c) Sensitivity analysis
(d) Influence diagrams

Q.11 Which of the following is NOT an objective of a risk audit?
(a) Confirming that risk management has been practiced throughout the project life cycle
(b) Confirming that the project is well managed and that the risks are being controlled
(c) Evaluating the effectiveness of risk responses in dealing with identified risks
(d) Ensuring that each risk identified and deemed critical has a computed expected value

Q.12 Arrange the following elements of the Project Cycle in the right order:
A. Project Appraisal
B. Feasibility Analysis
C. Execution
D. Project Selection
(a) A-B-C-D (b) B-A-C-D
(c) B-A-D-C (d) A-B-D-C

Q.13 You are the project manager in charge of developing a new shipping container for Globus Ocean Transport, which needs to withstand winds of 90 knots and swells of 30 meters. In determining the dimension of the container and the materials to be used in its fabrication, you convene a group of knowledgeable professionals to gather initial requirements, which will be included in the:

(a) Project charter
(b) Bill of materials
(c) WBS
(d) Project Statement of Work

Q.14 Feasibility report is prepared during the following phase of the project

(a) Intermediate phase
(b) Initial phase
(c) Final phase
(d) After completion of project

Q.15 The economic viability of a project can be assessed by the following methods

(a) Benefit cost ratio
(b) Pay back period and net present value
(c) Return on investment and internal rate of return
(d) All of these

Q.16 Break-even analysis can be used in

(a) Choosing the best product
(b) Selection among alternatives
(c) Choosing the critical method
(d) Selecting the best equipment

Q.17 A project on software development and construction of a bridge are similar to each other in terms of:

(a) Project Life-Cycle
(b) Project Management Process Groups
(c) Technology
(d) Resources

Q.18 Some of the accepted types of project appraisals are:

(a) Technical, Financial, Economical and Formal
(b) Technical, Financial, Ecological and Environmental
(c) Technical, Financial, Commercial and Environmental
(d) Commercial, Technical, Financial and Superficial

Q.19 The main purpose of conducting a project appraisal is to ensure the following:

(a) That the project has met all its objectives with respect to scope
(b) That the funds assigned for the project have not exceeded the initial assessment
(c) That all related areas of the project area examined before approving the execution of the project
(d) That the promoter is a qualified person who can be trusted with the project

Q.20 Technical appraisal of a project includes all of the following except

(a) Availability and transport of raw material
(b) Layout of the plant including employee safety
(c) Project scheduling and implementation
(d) Use of the latest technology on the project

Q.21 Demand forecasting techniques to ensure success of a proposed project include:

(a) Consumer surveys, Expert opinion, Delphi techniques, regression and trend analysis
(b) Consumer surveys, Expert opinion, Delphi techniques, regression and exit poll analysis
(c) Consumer analysis, Delphi surveys, Expert judgement, regression and trend studies
(d) Consumer studies, Delphi methods, Good judgement, TV and newspaper analysis

Q.22 The main reasons for conducting a financial appraisal are:

(a) Estimating the cost of the project and the ability of the Promoter to manage the project
(b) Obtaining funding for the project and estimating the promoter's ability to spend the money carefully
(c) Estimating the cost of the project and the break even point
(d) Estimating the cost of the project and the ability to obtain funding for this cost through debt or equity

Q.23 Risk planning on projects is mainly done to:

(a) Check if all technical options on the project have been explored

(b) Adopt a pro-active approach to ensure that the team members are prepared for both positive and negative risk events
(c) See whom to blame when things go wrong on the project
(d) Figure out how to get more money to fix things once problems happen on the project

Q.24 If the promoter of a project invests Rs. 25 lakhs and puts it in a bank deposit that offers 10% per year, how much money will he have at the end of 5 years?
(a) Rs. 27,50,000 (b) Rs. 41, 33,253
(c) Rs. 37,33,750 (d) Rs. 40,26,275

Q.25 Project A has an internal rate of return (IRR) of 21 percent in 2 years. Project B has an IRR of 17 percent in 3 years. Project C has an IRR of 31 percent in 2.5 years. Project D has an IRR of 29 percent in 1 year. Which of these would be the BEST project for implementation financially?
(a) Project A (b) Project D
(c) Project C (d) Project B

Q.26 Risk on a project involves:
(a) Pro-actively planning and preparing for positive and negative risk events as a one time process during the planning phase
(b) Re-actively ensuring that all positive and negative risk events are taken care off properly
(c) Pro-actively planning and preparing for positive and negative risk in an iterative manner throughout the project life
(d) Reactively checking to ensure that all new risk events do not impact the project negatively

Q.27 Risk is defined as:
(a) A negative event that may occur on a project
(b) A positive event that may occur on a project
(c) An event that may disturb the team on a project
(d) Both (a) and (b)

Q.28 The initial investment cash outflow by a company is Rs. 4,00,000 in a project and the cash inflows for the first five years are 80,000, 1,00,000, 50,000, 1,20,000 and 85,000. The IRR for this project will be
(a) 3.35% (b) 3.26%
(c) 3.12% (d) 2.79%

Q.29 The document that officially sanctions the project is the
(a) Project charter (b) Project plan
(c) Feasibility study (d) Cost-benefit analysis

Q.30 Project statement of work is done in
(a) Initiation Process (b) Planning Process
(c) Executing Process (d) Closing Process

Q.31 With respect to initiation process in project management, which are correct?
1. Chances of success is high
2. Risk is low
3. Costs are low
4. Stakeholders influences is high
(a) All of these (b) 1 and 2
(c) 2 and 3 (d) 3 and 4

Q.32 Which of the following is not a part of the project development process?
(a) Monitoring & control phase
(b) Pre-investment phase
(c) Operation phase
(d) Closing phase

Q.33 You as a project manager wants to do demand forecast for you new product. Which type of forecast will you do to analyse the market?
(a) Qualitative
(b) Quantitative
(c) Any one of the above
(d) Moving average method

Q.34 In below project selection methods which is not a quantitative method?
(a) Internal rate of return
(b) Opportunity cost
(c) Payback period
(d) Murder board

Q.35 Which of the project selection technique is most precise & conservative?
(a) NPV (b) IRR
(c) Opportunity cost (d) Payback period

Q.36 Which of the following two project selection techniques would give the same decision?

(a) NPV & payback period
(b) Payback Period & opportunity cost
(c) NPV & IRR
(d) IRR & Opportunity cost

Q.37 Which of the following characterises a risk?
1. Positive 2. Negative
3. Known 4. Unknown
(a) All of these (b) 1, 2 and 3
(c) 1 and 3 (d) 2 and 3

Q.38 In project identification technique of SWOT analysis factors internal to organisation are
1. Strength 2. Weakness
3. Opportunity 4. Threat
(a) 1 and 2 (b) 2 and 3
(c) 1 and 3 (d) 2 and 4

Q.39 Statement of work in a project has following elements:
1. Strategic plan
2. Stakeholders list including (PM & team)
3. scope
4. Business need
(a) All of these (b) 1, 3 and 4
(c) 1, 2 and 4 (d) 3 and 4

Q.40 The document containing information about identified project risks, security and solution is called
(a) Project charter (b) Statement of walk
(c) Log register (d) Risk register

Q.41 Risk identification and assessment is done :
(a) During the start of project
(b) At the end of project
(c) During the full project life cycle continually
(d) At the execution phase only

Q.42 Demand forecasting is attribute of which of the following appraisal.
(a) Technical appraisal
(b) Financial appraisal
(c) Market appraisal
(d) economical appraisal

Q.43 What are techniques used in market appraisal?
(i) Opinion Poll
(ii) Delphi technique
(iii) Trend analysis
(iv) Regression technique
(a) All of these (b) (i) and (ii)
(c) (ii) and (iii) (d) (iii) and (iv)

Q.44 In weighted moving average method of demand forecasting, more weight is given to the
(a) less recent data
(b) more recent data
(c) any one of (a), (b)
(d) all are given equal weight

Q.45 Financial Appraisal is done for
(i) Estimating Project Cost
(ii) Estimating Sales and production
(iii) To avoid crisis of cash liquidity
(iv) Estimation of funding
(a) All of above (b) (ii) and (iii)
(c) (i) and (iv) (d) None of these

Q.46 Which of the following appraisal is done to provide benefits to poorer sections of the society?
(a) Economical Appraisal
(b) Ecological Analysis
(c) Social Cost Benefit Analysis
(d) Technical Analysis

Q.47 If there is a scenario of unemployment in a region, then the shadow price of wage of the labour from its regular wage will be
(a) more (b) less
(c) equal to (d) can't be predicted

Q.48 Which of the following statements concerning the NPV is not true?
(a) The NPV technique takes account of the time value of money
(b) The NPV of a project is the sum of all the discounted cash flows associated with a project
(c) The NPV technique takes account of all the cash flows associated with a project
(d) If two competing projects are being considered, the one expected to yield the lowest NPV should be selected

Q.49 A risk-return trade-off function
(a) Shows the minimum expected return required to compensate an investor for accepting various levels of risk.
(b) Slopes upward for a risk averse decision maker

(c) Is horizontal for a risk neutral decision maker
(d) All of the above are correct

Q.50 In order to get a "feel" of the relationship between the product and its market, the project analyst may informally talk to customers, competitors, middlemen, and others in the industry. This is a ________analysis?
(a) Demand Analysis
(b) Situational Analysis
(c) Customer Analysis
(d) Product Analysis

Q.51 _______is the discounting rate which delivers a Net Present Value equal to zero?
(a) ARR (b) IRR
(c) NPV (d) Profitability Index

Q.52 Arrange the following as per the stages of Risk Management?
A. Analysis of probability and consequences
B. Risk identification : determine specific risk factors
C. Control and documentation
D. Risk mitigation strategies
(a) A-B-D-B (b) B-A-D-C
(c) B-D-A-C (d) B-A-C-D

Q.53 Arrange the steps in Decision tree Analysis?
A. Delineating the decision
B. Identifying the problem and alternatives-Imaginative efforts-risk and uncertainty
C. Evaluating various decision alternatives
D. Specifying probabilities and monetary outcomes
(a) B-C-D-A (b) B-A-D-C
(c) B-A-C-D (d) A-B-C-D

Q.54 _______is a time-phased budget that is used as a basis against which to measure, monitor, and control overall cost performance on the project?
(a) Cost Budget (b) Cost Baseline
(c) Cost input (d) Cost output

Q.55 Six Sigma is a business-driven, multi-dimensional structured approach to:
(a) Reducing process variability
(b) Increasing customer satisfaction
(c) Lowering Defects
(d) All of the above

Q.56 Choose the odd one with respect to Conflict Resolution?
(a) Mediate (b) Revenge
(c) Arbitrate (d) Control

Q.57 Project are particularly susceptible to risk because
(a) Murphy's law states that "if something can go wrong, it will"
(b) There is uncertainty in all projects
(c) Project management tools are generally unavailable at the project team level
(d) There are never enough resources to do the job

Q.58 Most statistical simulations of budgets, schedules, and resource allocations use which one of he following approaches?
(a) PERT
(b) Decision-tree analysis
(c) Present value analysis
(d) Monte Carlo analysis

Q.59 The simplest form of quantitative risk analysis and modeling techniques is
(a) Probability analysis
(b) Sensitivity analysis
(c) Delphi technique
(d) Utility theory

Q.60 What is the primary difference between a risk audit and a risk reassessment?
(a) A risk reassessment is conducted at the completion of a major phase; audits are conducted after the project is complete.
(b) Project stakeholders conduct risk audits; management conducts reassessments.
(c) Risk reassessments are regularly scheduled; risk audits are performed as defined in the project's risk management plan.
(d) There is no difference; they are virtually the same.

Answers

1. (c)	2. (b)	3. (c)	4. (b)	5. (d)
6. (a)	7. (c)	8. (d)	9. (c)	10. (a)
11. (d)	12. (c)	13. (a)	14. (b)	15. (d)
16. (b)	17. (b)	18. (c)	19. (c)	20. (c)

21. (a)	22. (d)	23. (b)	24. (d)	25. (c)
26. (c)	27. (d)	28. (d)	29. (a)	30. (a)
31. (d)	32. (d)	33. (a)	34. (d)	35. (a)
36. (c)	37. (a)	38. (a)	39. (b)	40. (d)
41. (c)	42. (c)	43. (a)	44. (b)	45. (a)
46. (c)	47. (b)	48. (d)	49. (d)	50. (b)
51. (b)	52. (b)	53. (b)	54. (b)	55. (d)
56. (b)	57. (b)	58. (d)	59. (b)	60. (c)

Explanations

1. (c)
Because Schedule Performance Index (SPI) and Cost Performance Index (CPI) are expressed as ratios, they can be used to show performance for a specific time period or trends over a longtime horizon.

2. (b)
A positive schedule variance indicates that the project is ahead of schedule. A positive cost variance indicates that the project has incurred less cost than estimated for the work accomplished; therefore, the project is under budget.

3. (c)
Preliminary project report contains topographic details and soil survey along alternate alignments, consideration of geometric design and other requirements of alignment, preparation of plans and comparison of alternate routes, economic analysis and selection of final alignment.

6. (a)
Scope, time, and cost are examples of project baselines to be part of the project management plan.

7. (c)
The Delphi technique provides a means for arriving at a consensus using a panel of experts to determine a solution to a specific problem. Project risk experts are identified but participate anonymously. Each panelist answers a questionnaire. Then the responses, along with opinions and justifications, are evaluated, and statistical feedback is given to each panel member. The process continues until group responses converge toward a solution.

8. (d)
It is often more effective to take early action to reduce probability and/or impact of a risk occurring on a project than attempting to repair the damage after the risk has occur.

9. (c)
Risk reassessment is an ongoing activity by the project team. Risks should be discussed at every status meeting. Risk audits are performed during the project life cycle to examine and document the effectiveness of risk responses. They are conducted at appropriate frequencies as defined in the risk management plan.

10. (a)
Perform qualitative risk analysis requires accurate and unbiased data. The use of low-quality data may result in a qualitative risk analysis that is of little use to the project manager regarding understanding of the risk, data available about the risk, data quality, and data reliability and integrity.

11. (d)
It is not feasible or necessary to quantify every risk. Therefore, a risk audit should never have as an objective to ensure that each project risk has a computed expected value.

13. (a)
The project charter documents include the business needs, assumptions, constraints, understanding of the customer needs and high-level requirements and what the new product, service, or result is to satisfy. It is the document used to formally authorize the project.

28. (d)
Net present value NPV)
= Present value of cash inflow – Present value of cash outflow
= 0

$$\frac{80000}{(1+r)}+\frac{100000}{(1+r)^2}+\frac{50000}{(1+r)^3}+\frac{120000}{(1+r)^4}+\frac{85000}{(1+r)^5}$$

$-400000 = 0$

$r = 0.02796$ or 2.796%

So, IRR for this project will be 2.79%

31. (d)
At the initiating stage :
1. Costs and low
2. Success chances are low
3. Risk in high
4. stakeholder influence is more.

34. (d)
It is a process where a committee asks the questions related to particular projects.

35. (a)
Net present value is the difference between projects current value of cash in flow and cash out flow. It take into consideration the future value of money.

38. (a)
Strength & weakness are always internal factor be it project or real life scenario.

39. (b)
Project manager & Team name are not in statement of work.

41. (c)
Risk of identified continually in project, it is ongoing process.

47. (b)
When there is unemployment then people will agree to work even on less wage than the normal so at least they would get some work.

48. (d)
If two competing projects are being considered, the one expected to yield the HIGHEST NPV should be selected.

49. (d)
The risk-return trade-off is the principle that potential return rises with an increase in risk. Low levels of uncertainty or risk are associated with low potential returns, whereas high levels of uncertainty or risk are associated with high potential returns. According to the risk-return trade-off, invested money can render higher profits only if the investor is willing to accept the possibility of losses.

57. (b)
Every project has uncertainty associated with it because a project by its definition is a temporary endeavor undertaken to create a unique product, service, or result. Risks may be known or unknown,

58. (d)
Simulations are typically performed using Monte Carlo in which a project model is computed many times with the input values chosen at random for each iteration from the probability distribution of these variables. Monte Carlo analysis supports various statistical distributions (normal, triangular, beta, uniform, etc.) used in estimating budgets, schedules, and resource allocations.

59. (b)
Sensitivity analysis, as a quantitative risk analysis and modeling technique, helps to determine the risks that have the most potential impact on the project. It examines the extent to which the uncertainty of each project element affects the objective being examined when all other uncertain elements are held at their baseline values.

60. (c)
Risk reassessments are regularly scheduled; risk audits arc performed as defined in (he project's risk management plan.
Risk reassessment is an ongoing activity by the project team. Risks should be discussed at every status meeting. Risk audits are performed during the project life cycle to examine and document the effectiveness of risk responses. They are conducted at appropriate frequencies as denned in the risk management plan.

■■■■

4 Financial Analysis and Project Finances

4.1 FINANCIAL ANALYSIS/FINANCIAL FEASIBILITY

Financial analysis refers to an activity of assessing financial statements to judge the financial performance of a company. It helps in assessing profitability, solvency, liquidity and stability.

Financial appraisal of project consists of two major areas *viz.*, arriving at the cost of the project and arriving at the appropriate means of financing the project.

Financial Appraisal Involves:

- Estimation of project cost
- Estimation of project operating cost
- Estimation of project funds
- Estimation of rate of return

Methods of Financial Analysis/Criteria Used for Selection of Investment Opportunities

1. Average Rate of Return on Investment or Accounting Rate of Return (ARR) or Return on Investment (ROI)

This is an accounting method. There is no agreement on the definition and a number of alternative methods of calculating it are available. The most common ratio is:

$$\text{ARR} = \frac{\text{Average Annual Profit after taxes}}{\text{Average Investment over the project life}} \times 100$$

- Average Annual Profit after taxes is calculated by adding up the after-tax profits for each year of project life and dividing it by the no. of years of estimated useful life (for annuities, after-tax profit is equal to one year's profit).
- Average Investment over the project life is computed by dividing the net investment by two (straight line depreciation is assumed) and adding the salvage value that would be received at the end of the project life (since it remains invested throughout) and full amount of working capital required.

  Average investment over project life

$$= \frac{1}{2}\,[\text{Initial cost} - \text{Salvage value}] + \text{Salvage value} + \text{Working capital}$$

EXAMPLE 4.1

Determine the ARR from the following data:

Sl. No.	Item Details	Machine M (Rs.)	Machine N (Rs.)
1.	Initial Purchase Price	60,000	60,000
2.	Yearly Income after depreciation & Income Tax		

— First Year	4,000	12,000
— Second Year	5,000	9,000
— Third Year	7,000	7,000
— Fourth Year	9,000	5,000
— Fifth Year	12,000	4,000
3. Estimated Life (years)	5	5
4. Estimated Salvage Value	3,000	3,000
5. Working Capital required	—	—

ARR can be worked out as below:

(straight line depreciation is assumed)

Average Income after taxes/year $\frac{37,000}{5}$ $\frac{37,000}{5}$

i.e., 7400   7400 ...(A)

Average Investment over the project Life = 1/2 (initial cost – Salvage Value) + Salvage Value + Working Capital

= 1/2 (60,000 – 3,000) + 3000 + 0

= 28,500 + 3,000 = 31,500 ...(B)

ARR $= \frac{7,400(A)}{31,500(B)} \times 100$

= 23.5%

Alternative Computation of ARR

Some analysis use the Initial Cost instead of the Average investment over the project life, in which case,

$$\text{ARR} = \frac{7,400}{60,000} \times 100 = 12.3\%$$

Companies select projects where ARR exceeds a 'predetermined value' or they rank the alternative proposals in the descending order of magnitude of ARR for selecting the most attractive one.

NOTE

Working capital = Current Assets – Current Liabilities

It represents operating liquidity available to a business.

If current assets are less than current liabilities then working capital is deficit.

2. Payback Period (PB)

- It is a traditional method and widely used for project evaluation.
- It is a measure in terms of time which will take to recover the cost of investment.
- If Cash in Flow After Tax (CFAT) per annum is uniform/even/constant, then

$$\text{Payback Period (PB)} = \frac{\text{Initial investment}}{\text{Constant annual cash flow (CFAT)}}$$

- If cash inflow is non even or variable year to year then

$$\text{Payback Period (PB)} = A + \frac{B}{C}$$

Where, A = Last period with negative cumulative cash outflow (Net invested cash flow)
B = Absolute value of cumulative cash outflow (i.e., Net invested cash flow) at the end of period A
C = Cash in flow during the next period after A

NOTE : Accept the project only if its Payback period is less than the targeted Payback period.

- If there are two projects then accept the project with least payback period.

EXAMPLE 4.2

MADE EASY is planning to undertake a project requiring initial investment of Rs. 105 crores. The project is expected to generate Cash Flow After Tax (CFAT) of Rs. 25 crores per year for 7 years, then calculate the payback period of the project.

Solution:

$$\text{Payback Period (PB)} = \frac{\text{Initial Investment}}{\text{Annual Cash Flow}} = \frac{105}{25} = 4.2 \text{ years}$$

EXAMPLE 4.3

A company is planning to take a project requiring initial investment of Rs. 50 crores and is expected to generate Rs. 10 crores cash flow (CFAT) in 1st year, Rs. 13 crores in 2nd year, Rs. 16 crores in 3rd years, Rs. 19 crores in 4th year and Rs. 22 crores in 5th year. Calculate the Payback period of the project.

Solution:

Year	Cash Inflow After Tax (CFAT) (Rs. in crores)	Cumulative Cash Outflow or Net Invested Cash (Rs. in crores)
0	—	–50
1	+10	–40
2	+13	–27
3	+16	–11
4	+19	0
5	+22	0

where, A = Last period with negative net invested cash or cumulative cash flow = 3
B = Absolute value of cumulative cash flow at the end of period A
= Rs. 11 crores
C = Total cash flow during the period after A, i.e., after 3rd year
= Rs. 19 crores

$$PB = 3 + \frac{11}{19} = 3.58 \text{ years}$$

Advantages of Payback Period:

1. Easy and simple to calculate.
2. It can be used to measure risk in project.
3. For companies facing liquidity problems, it provides a good ranking of projects that would return money early.

Disadvantages of Payback Period:

1. It does not take into account the time value of money. To remove this drawback, discounted cash flow method can be used.
2. It does not take into account the cash flows that occur after payback period.

3. Discounted Cash Flow Techniques

Traditional methods of project evaluation do not take into account the total benefits from the entire life cycle of a project and they do not consider the time value of money. The techniques described below, discount the cash flows by the cost of capital—a discounting factor for adjusting time value of money. This analysis uses the future free cash flow projections and discounts them to arrive at present value estimates which is used to assess the potential of investment.

Discounted cash flow is calculated as

$$\text{DCF} = \frac{CF_1}{(1+r)^1} + \frac{CF_2}{(1+r)^2} + \ldots \frac{CF_n}{(1+r)^n}$$

CF = Future cash flow, r = Discounted rate (WACC), 1, 2, ... n is period.

If DCF > Present value (PV)/Current cost, then project should be preferred.

4.2 PRESENT VALUE OF MONEY (THE CONCEPT OF DISCOUNTING)

We have seen that the future value of a given sum of money is calculated by the process of compounding. Similarly, the present value is calculated through the process of discounting. Discounting refers to the procedure of calculating the present value of future cash flows by applying suitable interest factor, thus adjusting for the time value of money. The technique of discounting is an exact opposite of the technique of compounding.

Present Value of Future Money Receivable after 'n' Years

The maturity value of Rs. 'P' at an interest rate of 'i %' p.a., compounded annually, for a period of 'n' years is given by

$$\text{Maturity value (MV)} = P(1 + i)^n$$

When the future value (i.e., Maturity value) is known, the present value can be found out form the above formula. Accordingly,

$$P = \frac{MV}{(1+i)^n}$$

EXAMPLE 4.4

An investor wants to get a sum of Rs. 4,00,000 after 5 years. The prevailing rate of interest for deposits is 11% p.a., and the interest rate is expected to remain stable. Interest is compounded annually. How much money should the investor invest now?

Solution:

$$P = \frac{MV}{(1+i)^n}$$

Substituting the given data,

$$P = \frac{4,00,000}{(1+0.11)^5}$$

$$P = 4,00,000 \times \left[\frac{1}{(1+0.11)^5}\right]$$

$$= 4,00,000 \times 0.5935 = \text{Rs. } 2,37,400$$

∴ Annual to be invested by the investor now = Rs. 2,37,400

Present Value of Perpetual Annuity (or Perpetuity)

When the periodical payments are received at annual intervals perpetually, the present value of such future cash flows is called the present value of perpetuity.

Let A = amount paid (or received) at the end of every year

I = interest rate per annum

$$\text{Present value of perpetual annuity} = A \times \left(\frac{1}{i}\right)$$

EXAMPLE 4.5

If a sum of Rs. 1500 is receivable every year perpetually, find out the present value of the perpetual cash flow. Assume a discount rate of 16% p.a.

Solution:

$$\text{Present value of perpetual annuity} = A \times \left(\frac{1}{i}\right) = 1500 \times \left(\frac{1}{0.16}\right) = \text{Rs. } 9375$$

4.3 FUTURE VALUE OF A SERIES OF CASH FLOW (FUTURE VALUE OF ANNUITY)

A series of payments of a fixed sum at regular intervals over a period of time is called Annuity. We come across investments of this nature in recurring deposits, pension/provident funds, life insurance policies etc., wherein the investor makes investments of constant amounts at regular time intervals, over a period of time:

The maturity value of Annuity (i.e., an annual series of investments) is given by the following relationship:

$$\text{Future value} = \text{Maturity value} = A\left[\frac{\{1+i\}^n - 1}{i}\right]$$

Where, A = Annuity amount (i.e., constant amount invested every year)

i = Interest rate

and n = Number of years

EXAMPLE 4.6

Calculate the compound value (i.e., maturity value) of Rs. 1000.00 invested every year for a period of 10 years, at an interest rate of 12% per annum.

Solution:

$$\text{Maturity value} = A\left[\frac{\{1+i\}^n - 1}{i}\right]$$

Substituting the given data,

$$\text{Maturity value} = 1000\left[\frac{\{1+0.12\}^{10} - 1}{12}\right]$$

$$= 1000\left[\frac{3.1058 - 1}{0.12}\right]$$

$$= 1000\left[\frac{2.1058}{0.12}\right]$$

$$= \text{Rs. } 17548.00$$

4.4 SINKING FUND PAYMENT

Sinking fund refers to the creation of a fund wherein money is accumulated by setting aside a certain sum of money periodically in such a way that the fund along with the interest accrued gets accumulated so a predetermined value at the end of a specified period of time.

Suppose I am in need of, say, Rs. 10 lakhs after 9 years from now. If I am prepared to invest a fixed sum of money regularly every year and if the money so invested is expected to earn interest at the rate of, say, 12% p.a., what shall the annual installments to be invested? The annual instalment is called the 'sinking fund'. This can be arrived at as under:

Expected Maturity value at the end of 9 years

= Rs. 1000000

Interest rate = 12% p.a.

Let the annual instalment to be invested = A

$$\text{Maturity value (MV)} = A\left[\frac{\{(1+i)\}^n - 1}{i}\right]$$

$$A = \frac{MV.i}{\left[(1+i)^n - 1\right]}$$

Substituting the given values,

$$A = \left[\frac{1000000 \times 0.12}{(1+0.12)^9 - 1}\right]$$

$$A = \left[\frac{120000}{(2.7731-1)}\right]$$

∴ Annual sinking fund to be invested = Rs. 67,678

(i) Net Present Value (NPV)

- The money received today is worth more than the sum received in the future i.e., it has time value. This occurs for three reasons.
  - (a) Potential for earning interest (cost of finance)
  - (b) Impact of inflation
  - (c) Effect of risk
- Net Present Value (NPV) computes the net present value of the cash inflows (CFAT) in each of the future year by discounting them suitably and then subtracting the cash outflow in each year, which in our example is Rs. 60,000 equal for each of the two machines M and N.

A. Present Value of Cash Inflows, year after year

Year	Machine M			Machine N		
	Annual CFAT (Rs.)	Present Value factor	Present value	Annual CFAT (Rs.)	Present factor	Present value
1.	16.000	0.909	14,544	24.000	0.909	21,816
2.	17,000	08.26	14,042	21.000	0.826	17.346
3.	19,000	0.751	14,269	19,000	0.751	14,269
4.	21,000	0.683	14,343	17,000	0.683	11.611
5.	24.000	0.621	14,904	16,000	0.621	9.936
A. Total present value of cash in flow			72,102			74.978
B. Present value of cash outflow			60,000			60,000
Net Present Value (A – B)			12,102			14.978

Capital Expenditure proposal with a positive NPV or which exceeds a predetermined NPV, can be selected for implementation. Alternatively, *the proposals can be ranked in the descending order of the NPV.*

EXAMPLE 4.7

An investment of Rs. 100 crores is to be made today. What is the value of investment after two years if the interest rate is 10%?

Solution:

$$\text{Value after 1 year} = 100 \times 1.1 = 110 \text{ crores}$$

$$\text{Value after 2 years} = 110 \times 1.1 = 121 \text{ crores}$$

$$F = P(1 + r)^n$$

F = Future value after n years P = Present value

n = Number of period r = Interest rate

Present value for future sum is, $P = \dfrac{F}{(1+r)^n}$

$$P = F(1 + r)^{-n}$$

$(1 + r)^{-n}$ is known as discount factor or present value factor

Net Present Value, NPV = [Present value of cash inflow] – [Present value of cash outflow (investment)]

Note	
• If NPV (Net Present Value) > 0, Then project is financially viable. = 0, Then project just meets breakeven. < 0, Then project is not financially viable. Project with increase NPV is preferred.	• Benefit Cost Ratio $BCR = \dfrac{\text{Total present value of benefits}}{\text{Total present value of investments}}$ $> 1 \Rightarrow$ Accept $< 1 \Rightarrow$ Reject $= 1 \Rightarrow$ Breakeven situation

EXAMPLE 4.8

Initial investment (cash outflow) is Rs. 2,40,000 and cash in flow (after tax) is Rs. 80,000 in 1st year, 1,20,000 in 2nd year, 70,000 in 3rd year, 40,000 in 4th year and 20,000 in 5th year. If the companies cost of capital is 9% (interest rate) then calculate Net Present Value (NPV).

Solution:

Year	Cash flow (Rs.)	D.F. at 9%	Present Value (Rs.)
1.	80,000	0.917	73,360
2.	1,20,000	0.842	1,01,040
3.	70,000	0.772	54,040
4.	40,000	0.708	28,320
5.	20,000	0.650	13,000
			Σ = 2,69,760

Present value of cash in flow, $A = 2,69,760$

Present value of cash outflow, $B = 2,40,000$

Net Present Value, $NPV = 2,69,760 - 2,4,0000 = +29760$

$$> 0$$

(ii) Internal Rate of Return (IRR)

- IRR method is know by many names i.e., *yield on investment method, marginal efficiency of capital method, marginal productivity of capital method, rate of return method, time adjusted rate of return method etc.* It also takes into account time value of money by discounting cash Inflows and cash Outflows.
- The Internal Rate of Return (IRR) also known as Economic Rate of Return (ERR) is a method of calculating rate of return. The term internal means that the calculation does not incorporate environmental factors (i.e., interest rate or inflation). It is also called the Discounted Cash Flow Rate of Return (DCFROR), sometimes it is also known as effective interest rate.

Definition: The IRR of an investment or project is the "Annualised effective or discount rate" that makes the net present value equal to zero.

OR

It is the discount rate at which the present value of all future cash flows is equal to the initial investment i.e., the rate at which investment achieve breakeven (NPV = 0).

OR

IRR of an investment is the discount rate at which the net present value of cost (negative cash flows) of the investment equals to the net present value of the benefits (positive cash flow) on the investment.

- IRR is used for capital budgeting and to compare the profitability of investments.
- Greater IRR are desirable. Project should be undertaken only when IRR > Cost of capital (Acceptable rate of return or interest rate).
- IRR give rate and is an indicator of efficiency/quality/yield of an investment where as NPV is an indicator of value of magnitude of investment.
- IRR can be used as an indicator to evaluate whether new plant is viable or extension of existing plant is viable.
- IRR can also be used in evaluating stock buy back programs. Sometimes companies own stock is better investment (higher IRR) than any other use of funds.

Let, r is IRR

$$\text{NPV} = \sum_{n=0}^{N}\frac{Cn}{(1+r)^n} - \text{Present cost outflow (initial cost)} = 0$$

n = no. of periods/years

Cn = Cash inflow in n^{th} period/year

EXAMPLE 4.9

Find IRR for the tabulated data shown below:

Year (n)	Cash flow (Cn)
0	–1,23,400
1	+36,200
2	+54,800
3	+48,100

– is Outflow
+ is Inflow

Solution:

$$\text{NPV} = \text{Present value of cash in flow} - \text{Present value of cash outflow} = 0$$

$$\frac{36{,}200}{(1+r)^1}+\frac{54{,}800}{(1+r)^2}+\frac{48{,}100}{(1+r)^3}-1{,}23{,}400=0$$

$$\Rightarrow \qquad r = 0.0596 = 5.96\%$$

Projects are selected if the IRR is higher than the rate of return predetermined by the company.

Important Terms :

- *Total cost of production = Cost of production + Total administrative expenses + Total sales expenses + Royalty, know-how payable*
- *Operating profits = Gross profit before interest and financial expenses – Total financial expenses – Depreciation charges*
- *Profit before Taxation = Operating profits + Cost of production – Preliminary expenses written off*
- *Profit after Tax = Provision for Taxation – Profit before Taxation*
- *Net Cash Accruals = Retained profits + Depreciation charges + Preliminary expenses written off*

4.5 PROJECT FINANCES

We use combination of equity and debt to finance the project. The proper equity-debt combination for a project depends upon the revenue earning capacity of the project. Though equity does not attract any interest and does not involve any repayment obligation, the debt obtained for financing the project is to be repaid along with interest. Hence, the revenue earning capacity of the project should be sufficient enough to meet the repayment of debt with interest. Moreover, the project should also generate additional revenue to meet the cost of equity capital. Though equity is contributed by the project promoters, it is not without any cost. An entrepreneur invests his capital in a project with a hope of getting returns on his capital invested apart from meeting debt obligations.

4.6 CAPITAL STRUCTURE OF PROJECT

The capital structure is the particular combination of debt and equity used by a company to finance its overall operations and growth. Debt comes in the form of bond issues or loans, while equity may come in the form of common stock, preferred stock, or retained earnings. Short-term debt such as working capital requirements is also considered to be part of the capital structure.

Debt consists of borrowed money that is due back to the lender, commonly with interest expense.

Equity consists of ownership rights in the company, without the need to pay back any investment.

Difference between Debt and Equity

Feature	Debt	Equity
Capital	Protected	Can go down
Upside	Generally no upside	Upside is possible
Fixed interest payment	Present	Nothing is guaranteed
Example	Fixed deposits, FMPs, POMIS	Shares, ELSS Mutual funds, Index ETFs.

4.7 DEBT INSTRUMENT

A debt instrument is a fixed income asset that allows the lender (or giver) to earn a fixed interest on it besides getting the principal back while the issuer (or taker) can use it to raise funds at a cost. Debt acts as a legal obligation on the issuer (or taker) part to repay the borrowed sum along with interest to the lender on a timely basis. A debt instrument can be in paper or electronic form. Bonds, debentures, leases, certificates, bills of exchange and promissory notes are examples of debt instruments.

These instruments also give market participants the option to transfer the ownership of debt obligation from one party to another. The lender receives a fixed amount of interest during the lifetime of the instrument.

Debt instruments provide fixed and higher returns, thus giving them an edge over bank fixed deposits. The duration of debt instruments can either be long-term or short-term. Funds raised through short-term debt instruments are to be repaid within a year.

However, long-term debt instruments are the ones that are paid over a year or more. Credit card bills and treasury notes are examples of short-term debt whereas long-term loans and mortgages form part of long-term debt instruments.

Some of the common types of the debt instrument are:

1. Debentures
2. Bond
3. Mortgage
4. Treasury Bills

1. Debentures

Debentures are not backed by any security. They are issued by the company to raise medium and long term funds. They form the part of the capital structure of the company, reflect on the balance sheet but are not clubbed with the share capital.

The company issues the debentures under the seal of the company. The document of debenture contains the information regarding the rate of interest, repayment amount, and also the time when the company has to repay the debenture. The company can issue various types of debentures. They can be convertible or non-convertible, redeemable or non-redeemable, etc.

Convertibility: Shares cannot be converted into debentures whereas debentures can be converted into shares if the terms of issue so provide, and in that case these are known as convertible debentures.

From the Point of view of Tenure:

(a) **Redeemable Debentures:** Redeemable debentures are those which are payable on the expiry of the specific period either in lump sum or in Instalments during the life time of the company. Debentures can be redeemed either at par or at premium.

(b) **Irredeemable Debentures:** Irredeemable debentures are also known as Perpetual Debentures because the company does not give any undertaking for the repayment of money borrowed by issuing such debentures. These debentures are repayable on the winding-up of a company or on the expiry of a long period.

Terms of Issue of Debentures: Depending upon the terms and conditions of issue and redemption of debentures, the following six situations are commonly found in practice.

(i) Issued at par and redeemable at par
(ii) Issued at discount and redeemable at par
(iii) Issued at a premium and redeemable at par
(iv) Issued at par and redeemable at a premium
(v) Issued at a discount and redeemable at a premium
(vi) Issued at a premium and redeemable at a premium

2. Bonds

Bonds on the other hands are issued generally by the government, central bank or large companies are backed by a security. Bonds also ensure payment of fixed interest rates to the lenders of the money. On maturity of the bond, the principal amount is paid back. Bonds essentially work the way loans do.

Note: These days, bonds are also being issued by semi-government and non-governmental organisations. The terms 'debentures' and 'Bonds' are now being used inter-changeably.

3. Mortgage

A mortgage is a loan against a residential property. It is secured by an associated property. In a case of failure of payment, the property can be seized and sold to recover the loaned amount.

4. Treasury Bills

Treasury bills are short-term debt instruments that mature within a year. They can be redeemed only at maturity. They are sold at a discount if sold before maturity.

Share: Shares are units of ownership interest in a corporation or financial asset that provide for an equal distribution in any profits, if any are declared, in the form of dividends.

The share capital of a company is of two types:Preferential Share Capital and Equity Share Capital

Preferential Share Capital: The preferential share capital is that part of the Issued share capital of the company carrying a preferential right for:

- **Dividend Payment:** A fixed amount or amount calculated at a fixed rate. This might/might not be subject to income tax.
- **Repayment:** In case of a winding up or repayment of the amount of paid-up share capital, there is a preferential right to the payment of any fixed premium or premium on any fixed scale. The Memorandum or Articles of the company specifies the same.\

Equity Share Capital (Equity Shares): All share capital which is NOT preferential share capital is Equity Share Capital.

Authorized and Issued Shares: Authorized shares comprise the number of shares a company's board of directors may issue. Issued shares comprise the number of shares that are given to shareholders and counted for purposes of ownership.

"Paid up capital : It is the actual amount paid up by the investors when they buy the shares."

Debt-Equity Ratio: Savvy companies have learned to incorporate both debt and equity into their corporate strategies. At times, however, companies may rely too heavily on external funding, and debt in particular.The Debt-to-Equity (D/E) ratio is useful in determining the riskiness of a company's borrowing practices.Usually, a company that is heavily financed by debt has a more aggressive capital structure and therefore poses greater risk to investors.

Companies benefit from debt because of its tax advantages; interest payments made as a result of borrowing funds may be tax deductible. Debt also allows a company or business to retain ownership, unlike equity. Additionally, in times of low interest rates, debt is abundant and easy to access.

Equity allows outside investors to take partial ownership in the company. Equity is more expensive than debt, especially when interest rates are low. However, unlike debt, equity does not need to be paid back. This is a benefit to the company in the case of declining earnings. On the other hand, equity represents a claim by the owner on the future earnings of the company.

4.8 COST OF CAPITAL

Cost of Debt : The cost of debt measure is helpful in understanding the overall rate being paid by a company to use different types of debt financing which may include debt such as bonds or loans, among other types.

To calculate the cost of debt, a company must determine the total amount of interest it is paying on each of its debts for the year. Then it divides this number by the total of all of its debt. The result is the cost of debt.

The cost of debt formula is the effective interest rate multiplied by (1 - tax rate). The effective tax rate is the weighted average interest rate of a company's debt.

The rationale behind this calculation is based on the tax savings the company receives from claiming its interest as a business expense.

EXAMPLE 4.10

A company has a 10 million loan with a 5% interest rate and a 2 million loan with a 6% rate.If the tax rate is at 30% calculate the cost of debt.

Solution:

The effective interest rate on its debt is 5.2%.

The company's tax rate is 30%.

Thus, its cost of debt is 3.64% or 5.2% × (1-30%).

Cost of Debt Associated with Debenture

Cost of debt associated with debenture is of two types:

1. **Cost of irredeemable debt or perpetual debt:** Perpetual debt refers to issue of debentures which will not be redeemed during the lifetime of the company. We use the following formula to calculate the cost of redeemable debt or perpetual debt.

$$K_d = \frac{I}{SV} \times 100$$

Where,
K_d = Cost of debt
I = Interest payble
SV = Net proceeds from issue

Calculation of SV (to be used for calculation of both redeemable and irredeemable debt)

(a) When debentures are issued at par SV = Par Value – Floatation Charges

(b) When debentures are issued at discount SV = Par Value – Discount - Floatation Charges

(c) When debentures are issued at Premium SV = Par Value + Premium – Floatation Charges

NOTE

1. Floatation Charges include Underwriting commission, brokerage, printing expenses etc.
2. Suppose par value of one debenture is Rs.100, it means that the company would receive Rs. 100 by selling 1 debenture. While issuing debenture, the company may incur floatation charges or company may give discount. Because, floatation charges and discount are our expenses, it should be deducted from par value.
3. In the same way, premium is added to our proceeds because it's a gain for the company.

2. **Cost of Redeemable Debenture:** Redeemable debentures are payable after a specified duration. These debentures could also be issued at par, discount or premium.

(a) When debentures are issued at per and redeemable at par with floatation cost.

$$K_d = \frac{I + \frac{f}{n}}{\frac{SV + RV}{2}} \times 100$$

(b) Whend debentures are issued at premium or discount but redeemable at par with floatation cost.

$$K_d = \frac{I + \frac{f}{n} + \frac{d}{n} - \frac{pi}{n}}{\frac{SV + RV}{2}} \times 100$$

(c) When debentures are issued at par premium or discount but redeemable at premium weith floatation cost

$$K_d = \frac{I + \frac{f}{n} + \frac{d}{n} - \frac{pi}{n} + \frac{Pr}{n}}{\frac{SV + RV}{2}} \times 100$$

where,
K_d = Cost of debt
I = Interest payable
SV = Net proceeds from issue
RV = Redemption value
f = Floatation cost
d = Discount
pi = Premium on issue
Pr = Premium on redemption
n = Number of years

Cost of Equity: A firm's cost of equity represents the compensation the market demands in exchange for owning the asset and bearing the risk of ownership.

The traditional formula for the cost of equity is the dividend capitalization model and the capital asset pricing model (CAPM).

4.9 COST OF EQUITY FORMULA

Using the dividend capitalization model, the cost of equity is:

$$\text{Cost of equity} = \frac{DPS}{CMV} + GRD$$

where,
DPS = Dividends per share, for next year
CMV = current market value of stock
GRD = growth rate of dividends

Using the capital asset pricing model (CAPM),

Cost of Equity = Risk-Free Rate of Return + Beta × (Market Rate of Return – Risk-Free Rate of Return)

The risk-free rate is the rate of return paid on risk-free investments such as Treasuries. Beta is a measure of risk calculated as a regression on the company's stock price. The higher the volatility, the higher the beta and relative risk compared to the general market.In general, a company with a high beta, that is, a company with a high degree of risk will have a higher cost of equity.

EXAMPLE 4.11

The market price per share of a company is '125. The dividend per share (DPS) expected a year is '12 and DPS is expected to grow at a constant rate of 8% per annum. What will be the cost of the equity capital to the company?

Solution:

$$\text{Cost of equity} = \frac{DPS}{MPS} + r = \frac{12}{125} + 0.08$$

$$= 0.096 + 0.08 = 0.176$$

Thus, cost of equity = 17.6%

EXAMPLE 4.12

Imagine an investor is contemplating a stock worth Rs.100 per share today that pays a 3% annual dividend. The stock has a beta compared to the market of 1.3, Also, assume that the risk-free rate is 3% and this investor expects the market to rise in value by 8% per year.Find the cost of equity.

Solution:

The cost of equity based on the CAPM formula is 9.5%:

$$9.5\% = 3\% + 1.3 \times (8\% - 3\%)$$

Note: The cost of debt is generally lower than cost of equity.

Cost of Capital

The cost of capital is the total cost of raising capital, taking into account both the cost of equity and the cost of debt. A stable, well-performing company, will generally have a lower cost of capital. To calculate the cost of capital, the cost of equity and cost of debt must be weighted and then added together.

Emerging Sources of Funding : Angel Investor : An angel investor (also known as a private investor, seed investor or angel funder) is a high net worth individual who provides financial backing for small startups or entrepreneurs, typically in exchange for ownership equity in the company. It is noteworthy to appreciate that angle investor invests during the early stage of a project.

Crowd Funding : It is innovating way of raising money by launching product through the internet.

4.10 DEPRECIATION OF ASSETS

Definition and concept: In Accounting or Accountancy, the concept of Depreciation can be used to explain the loss of value of an asset used by a business over time. This is done by allocating the cost of the asset or item over the period that it is being used. One of the main reasons for using this concept, apart from reflecting costs related to the asset in the balance sheet of the business, is to reduce the taxable income or net income for the company. This is done by reducing the gross income of a company during a certain time period by the expense of the assets used during the same time period. This cost is called the depreciated cost and the asset is called a depreciable asset. Depreciation expense begins when the asset is put into service by the company for its business.

Calculation of the depreciation expense or value of an asset requires four inputs:

- Purchase value of the asset
- Salvage value (or remaining or residual value) of the asset at the end of its useful life
- Estimated useful life of the asset
- Method used to calculate depreciation

4.10.1 Types of Depreciation

A common classification of depreciation is as follows:

(a) Physical depreciation

(b) Functional depreciation

(c) Contingent depreciation

(a) Physical Depreciation: Depreciation resulting in physical impairment of an asset is known as physical depreciation. This results in lowering the ability of the asset to render its intended service. The primary cause of physical depreciation is wear and tear because of its constant use such as abrasion, shocks, vibration, impact etc. and the deterioration due to action of elements such as corrosion of pipe, chemical decomposition.

(b) Functional Depreciation: Functional depreciation often called OBSOLESCENCE and is defined as the loss in the value of the property due to change in fashion, design or structure or due to inadequacy to meet the growing demand, necessity of replacement due to new invention being more economical and more efficient etc.

(c) Contingent Depreciation: It is due to:

1. accidents (due to negligence)
2. disease (pollution of water, parasites)
3. diminution of supply (natural gas, electricity water etc.

Table: Difference between Depreciation and Obsolescence

Depreciation	Obsolescence
1. This is the physical loss in the value of property due to wear and tear and decay etc.	1. This is the functional loss in the value of the property due to change in design, structure, fashion, utility, demand etc.
2. It depends upon its original condition, quality of maintenance and mode of use.	2. It depends upon the technological advancement, art etc.
3. It varies with age.	3. Not dependent on age.
4. There are several methods for calculating the amount of depreciation.	4. No method is available for calculation of obsolescence.

4.10.2 Salvage Value (or Resale Value)

It is the value of the property at the end of its utility period without being dismantled. Salvage value implies that the property has further utility. A structure that has been well maintained and carefully handled will fetch more salvage value.

Scrap Value: The value of a property realized when it become absolutely useless except for sale as junk is its scrap value. The utility of the article is assumed to be zero.

Book Value: It is defined as the value of the property shown in the account books in that particular year i.e. the original cost less total depreciation till that year.

Methods for Calculating Depreciation: There are several methods of calculating depreciation. Some of them are simple and used regularly whereas some are relatively uncommon. The following notations have been used:

C_i = Initial cost of an asset at zero time (This will include the cost of asset +transportation charge + installation and other charges spent initially)

C_S = Salvage value (or scrape value) to be estimated at the end of utility period.

n = Life of the asset.

B_m = Book value at the end of period 'm'

4.10.3 Straight Line Method

In this method, the property is assumed to lose value by a constant amount every year. At the end of the life, the salvage value (or scrap value) left.

$$D_m = \frac{C_i - C_S}{n}$$

$$D_m = D_1 = D_2 = D_3 = D_n$$

$$B_m = C_1 - m\left(\frac{C_i - C_s}{n}\right)$$

DO YOU KNOW?

- This method is recommended for all the equipments/assets with constant demand and which do not face any obsolescence during their useful life. It is widely used in the case of all civil engineering appliances and applications.

4.11 DECLINING BALANCE METHOD (OR CONSTANT PERCENTAGE METHOD)

In this method, the property is assumed to lose value annually at a constant percentage of its book value.

Let FDB be the fixed percentage

$$D_1 = \text{Depreciation for } 1^{st} \text{ year} = C_i \times \text{FDB}$$

$$\therefore \quad B_1 = \text{Book value at end of } 1^{st} \text{ year}$$

$$= C_i - C_i \times \text{FDB} = C_i(1 - \text{FDB})$$

$$D_2 = \text{Depreciation for } 2^{nd} \text{ year}$$

$$= \text{Book value at beginning of } 2^{nd} \text{ year} \times \text{FDB}$$

$$= C_i(1 - \text{FDB})\,\text{FDB}$$

(Book value at beginning of 2nd = Book value at the end of 1st year) depreciation for second year.

$$\therefore \quad B_2 = C_i(1 - \text{FDB}) - C_i(1 - \text{FDB})\text{FDB}$$

$$= C_i(1 - \text{FDB})^2$$

Similarly, $B_3 = C_i(1 - \text{FDB})^3$

....

....

$\therefore$ Book value at the end of n years $= C_i(1 - \text{FDB})^n$

$$\therefore \quad C_S = C_i(1 - \text{FDB})^n$$

$$\therefore \quad \text{FDB} = 1 - \left(\frac{C_S}{C_i}\right)^{1/n}$$

DO YOU KNOW?

- In some cases, the cost of maintenance increases towards the end of the life, it may be desirable to charge large amount of depreciation in the early period and less in the fag period, so that the combined effect of depreciation and maintenance costs roughly remains the same.

This method is most suitable when the asset fears to be obsolete before its life such as electronic equipments etc., but there are two difficulties in using this method. The first difficulty is that this formula can never depreciate to its salvage value. But this is not a serious difficulty. In actual practice, the depreciation for the last year is taken as book value at the beginning of the last year minus salvage value. The other serious difficulty is that this method can not be used when the salvage value of the asset is zero.

Double Declining Balance Method: In this method also, the property is assumed to lose value annually by fixed factor of the book value.

FDDB = Fixed factor for double declining balance method.

FDDB is taken as double the straight line

Rate. i.e. $$\text{FDDB} = \frac{2}{n}$$

The process of calculation of depreciation and book value of each year is the same as adopted in the Declining Balance Method.

DO YOU KNOW?

- The applicability of the method is also similar i.e. when it is intended to charge larger portion of depreciation in the early life of the property, particularly the asset/equipment faces the obsolescence before the estimated life.

4.12 SUM OF THE YEARS DIGIT METHOD

This method also falls in the category of accelerated type depreciation method like Declining Balance Method and Double Declining Balance Method in which depreciation models assume that the value of an asset decreases at a decreasing rate.

In this method, the digits corresponding to the number of each years of life are listed in reverse order. The sum of these digits is then determined. The depreciation factor for any year is the reverse digit for the year divided by the sum of the digits, and depreciation for a year is the product of depreciation factor of that year and the amount ($C_i - C_S$). The general expression for the annual depreciation for any year (m) when the life is n years is expressed as

$$D_m = (C_i - C_S)\left[\frac{(n-m+1)}{\frac{n(n+1)}{2}}\right]$$

DO YOU KNOW?

- This method also provides rapid depreciation during the early years of life like the previous two methods but this method has one advantage over them that it enables properties to be depreciated to zero value and is easier to use in practice.

4.13 SINKING FUND METHOD

Sinking fund depreciation model assumes that the value of an asset decreases at an increasing rate. Equal amount (D) is assumed to be deposited into a sinking fund at the end of each year of the assets life. The sinking fund is ordinarily compounded annually, at the end of the estimated life, the amount accumulated equals the total depreciation of the asset (C_i- C_S). It is to be clearly understood that the depreciation amount is the sum of two components. The first component is the amount deposited into sinking fund and the second component is the amount of interest earned on the accumulated value of the sinking fund at the beginning of the particular year in question.

Let 'i' be the rate of interest. The first component of depreciation which is to be deposited equally at the end of each year

$$= (C_i - C_S)\left[\frac{A}{F}, i, n\right]$$

$$D = [C_i - C_S]\left[\frac{i}{(1+i)^n - 1}\right]$$

Combining both factors into account, the depreciation for m^{th} year by this method is enumerated as

$$D_m = D[1+i]^{m-1}$$
$$m = 1, 2, 3.....$$
$$D_1 = D$$
$$D_2 = D(1+i)$$
$$D_3 = D(1+i)^2$$
$$D_m = ...$$

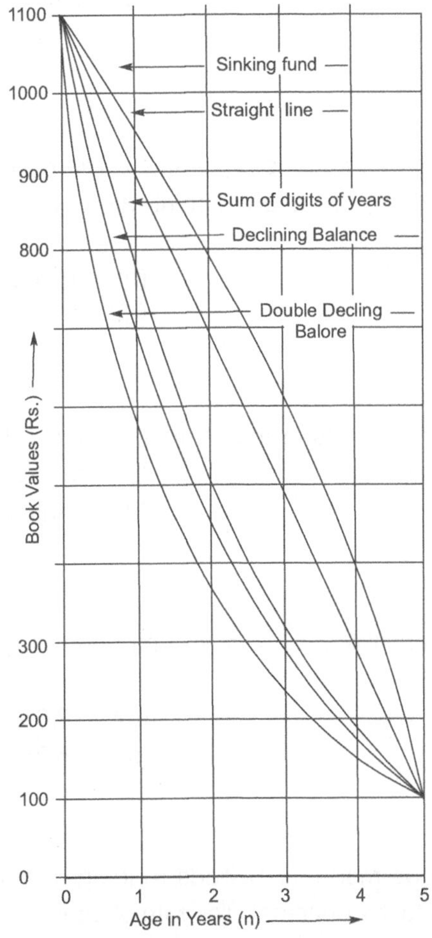

Fig. *Sinking Fund Method*

4.14 LIFE CYCLE COST (LCC)

It is also known as the 'Cradle to Grave Cost'. It is the cost of an item from its acquisition till it is abandoned after completion of its useful life period. The components of LCC are:

1. Cost of Acquisition

2. Cost of Operation
3. Cost of Repairs and Maintenance
4. Cost of Replacement of parts/components
5. Salvage Value

LCC = Cost of [Acquisition + Operation + Repair and Maintenance + replacement of components] – Salvage Value

LCC = Sum of items 1 to 4 above less Salvage Value.

In assessing the Value Index, if LCC is used

LCC = (Present value of all cost incurred) – (Present value of Salvage Value)

Previous ESE Prelims Questions

Q.1 During an assessment of economic viability of the project, the ratio of average annual earnings after tax to the average book investment after depreciation is called

(a) Benefit-Cost Ratio (BCR)
(b) Net Present Value (NPV)
(c) Pay-Back Period (PBP)
(d) Return on Investment (ROI)

[ESE-2020]

Ans. (d)

Q.2 The market price per share of a company is ₹125. The dividend per share (DPS) expected a year is ₹12 and DPS is expected to grow at a constant rate of 8% per annum. The cost of the equity capital to the company will be

(a) 17.6%
(b) 15.4%
(c) 13.2%
(d) 11.8%

[ESE-2020]

Ans. (a)

Q.3 An individual investor who invests in the e-project usually during an early stage is

(a) corporate strategic investor
(b) founder capital
(c) angel investor
(d) venture capital

[ESE-2020]

Ans. (c)

Q.4 If the nominal rate of interest is 12% and is compounded quarterly, the effective rate of interest per annum will be nearly

(a) 10.8%
(b) 12.6%
(c) 14.4%
(d) 16.2%

[ESE-2020]

Ans. (b)

Q.5 In a bank, deposits can be made for periods ranging from 6 months to 10 years. Every quarter, an interest will be added on to the principal. The rate of interest applied is 9% per annum for periods from 12 months to 23 months and 10% per annum for periods from 24 months to 120 months. An amount of ₹1,000 invested for 2 years to grow, will be nearly

(a) ₹1,218
(b) ₹1,334
(c) ₹1,414
(d) ₹1,538

[ESE-2020]

Ans. (a)

Q.6 A company has issued ₹20 million worth of non-convertible debentures, each at a face value of ₹100 at the rate of 12%. Each debenture is redeemable at a premium of 5%, after 10 years. If the net amount realized is ₹95 and tax rate is 40%, the cost per debenture will be

(a) 5.8% (b) 6.6%
(c) 7.4% (d) 8.2%

[ESE-2020]

Ans. (d)

Q.7 **Statement (I):** Raw materials are taken as traded items and their values at domestic and world prices are estimated.

Statement (II): Raw materials, which have a high value-to-volume ratio and involve proportionately high transport cost and are imported, are regarded as non-traded items.

(a) Both Statement (I) and Statement (II) are individually true and Statement (II) is the correct explanation of Statement (I)
(b) Both Statement (I) and Statement (II) are individually true but Statement (II) is not the correct explanation of Statement (I)
(c) Statement (I) is true but Statement (II) is false
(d) Statement (I) is false but Statement (II) is true

[ESE-2020]

Ans. (c)

Q.8 The creative technique applied when the available and required inputs as well as the desired outputs are listed, is

(a) Attribute listing (b) Direct dreaming
(c) Black box (d) Delphi

[ESE-2020]

Ans. (a)

Attribute listing is a means of getting you to focus on as many attributes of a product or problem as possible. In breaking down the elements of a problem or object, you can look at each in turn and generate new ideas. The technique is particularly useful for considering complex products or processes in that it allows you to consider each feature or stage and look at the associated attributes in detail.

Q.9 Under which one of the following circumstances is the project accepted as worthwhile, keeping the principal non-discounting criteria?

(a) The payback period (PBP) > target period
(b) The payback period (PBP) < target period
(c) The payback period (PBP) = target period
(d) The payback period (PBP) = 0

[ESE-2021]

Ans. (b)

Payback Period refers to the amount of time it takes to recover the cost of an investment. If it is less than the target period, investment is accepted as worthwhile. The lesser the Payback Period the more preferred is the investment.

Q.10 Consider the following statements regarding production:

The major aspects of production that may lead to sickness are

1. Increase in the cost of production.
2. Decrease in the quantity of production.
3. Quality of product not meeting the standards/customer expectation.

4. Producing more quantity than can be sold, leading to accumulation of stock.

Which of the above statements are correct?

(a) 1 and 2 only
(b) 1, 2, 3 and 4
(c) 2 and 3 only
(d) 1, 3 and 4 only

[ESE : 2022]

Ans. (b)

The major causes that may lead to sickness in production are insufficient resources and improper production and marketing policies. These causes lead to aspects such as

1. Increase in the cost of production.
2. Decrease in the quantity of production.
3. Quality of product not meeting the standards/customer expectation.
4. Producing more quantity than can be sold, leading to accumulation of stock.

Q.11 What are the factors that contribute to decline in unit cost with respect to the accumulated volume of production?

(a) Pioneering stage, Rapid growth stage and Economies of scale stage
(b) Learning effects, Technological improvements and Economies of scale
(c) Technological improvements stage, Maturity stage and Decline stage
(d) Pioneering stage, Rapid growth stage and Decline stage

[ESE : 2024]

Ans. (a)

■■■■

5 PLANNING PROCESS : Project Planning and Scheduling

5.1 PLANNING

In this phase you identify the general requirements of a project, break it down into discrete activities, determine the sequence and interdependence of these activities, estimate the resource requirements and find the critical path.

5.2 PROJECT MANAGEMENT PLAN

- It is a written document.
- It contains all the subsidiary plans needed for the project.
- It is created by the PM .
- Inputs are given from stakeholders/customers.
- The project management plan integrates all of the nine Knowledge Areas i.e. slope, time, cost, quality, HR, communication, Risk, pronouncement & project integration management.
- Though all the areas are not equally utilised in all projects. It all depends on the project manager and team that which area is given priority and which not.
- In project management plan following input elements are present :
  - Technique to be used
  - Project charter

Need Project Management Plan

- To get the required process & their interactions with other processes.
- To get the sequence of different Processes.
- To monitor & control processes & changes.
- To measure the project progress from baseline.
- To address the risk & issues.

Note : Project Manager & Team members personnel detail is not given in PMP. It can only tell the quantitative staff required & their cost.

5.3 DEVELOP PROJECT MANAGEMENT PLAN

- Inputs
  - Project Charter
  - Outputs from other processes
  - Enterprise environmental factors
  - Organizational process assets

- Tools & Techniques
  - Expert judgment
  - Facilitations techniques
- Outputs
  - Project management plan

Enterprise Environmental Factors

The enterprise environmental factors that can influence the Develop Project Charter process include, but are not limited to :

- Governmental standards, industry standards, or regulations (e.g. codes of conduct, quality standards, or worker protection standards).
- Organizational culture and structure, and
- Marketplace conditions.

Organizational Process Assets

The organizational process assets that can influence the Develop Project Charter process include, but are not limited to :

- Organizational standard processes, policies, and process definitions.
- Templates (e.g. project charter template). and
- Historical information and lessons learned knowledge base (e.g. project, records, and documents; all project closure information and documentation; information about both the results of previous project selection decisions and previous project performance information; and information from the risk management activity).

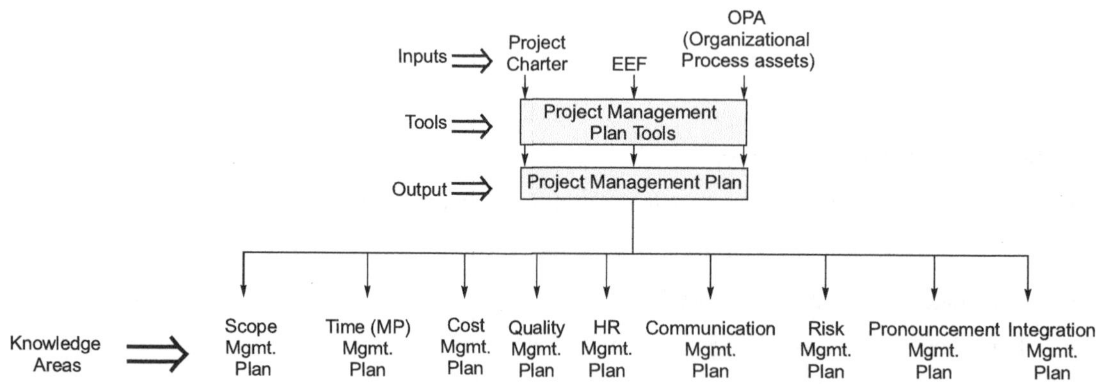

Project Management Plan

Core processes are usually interrelated and intertwined with each other and need to be performed in the same order on the majority of projects. They include three essential procedures for developing the project plan:

Planning and Defining the Scope: The first thing to do is to develop a clear scope statement as the project's foundation. The whole project will be built upon this and after that you can subdivide the major project objectives into clear and manageable deliverables.

Developing the Schedule: This requires a series of several procedures in order to create the project schedule that involves: Identifying and cataloguing the activities that must be performed to produce the various project deliverables, analyzing the sequence of activities and documenting any interactivity dependencies. Estimating all the work periods needed and how long it will take to complete individual activities. It is also useful to design a Milestone List, which lists the most important deliverables as way points, which will help you to evaluate the progress of a project.

Resource Planning: This includes the determination of resources (people, equipment, materials, etc.) and the quantities necessary to carry out the project activities, the development of a cost approximation of the resources required to complete the project and the cost budgeting and the allocation of the overall cost estimation to individual work packages.

Detailed Project Report: *DPR is nothing but a very detailed and elaborate plan for project, indicating the overall programme, different roles and responsibilities and activities and resources required. Success of a project is largely dependent on the quality of DPR.*

A DPR is a final detailed appraisal report on the project and a blueprint for its execution.

5.4 KICK OFF MEETING

- Once the project has been selected, the next step is for senior management to choose a Project Manager (PM).
- It is the Project Manger's job to make sure that project is properly planned, implemented and completed. PMs are some time chosen prior to the project selection also.
- The PM and Senior Manager call a meeting of people who are competent to begin planning.

A kick off meeting is the first meeting with the project team and the client of the project. This meeting would follow definition of the base elements for the project and other project planning activities. This meeting introduces the members of the project team and the client and provides the opportunity to discuss the role of the team member.

By displaying thorough knowledge of the goal & steps on how to reach it, the customer gains confidence in the team ability to deliver the work. Kick off means that the work starts.

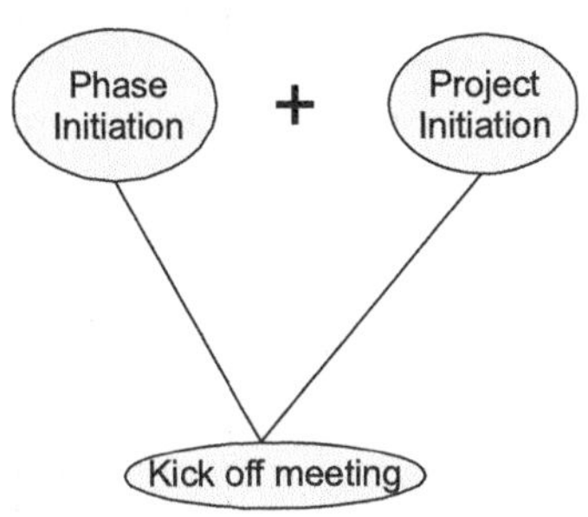

5.5 SCOPE OF PROJECT

When it comes to project planning, defining the project scope is the most critical step. In case if you start the project without knowing what you are supposed to be delivering at the end to the client and what the boundaries of the project are, there is little chance for you to success.

In the project scope definition, the elements within the scope and out of scope are well defined in order to clearly understand what will be the area under the project control. Therefore, you should identify more elements in detailed manner and divide them among the scope and out of scope.

Project scope is the work that must be done in order to deliver a product according to the product's scope (required functions and features).

Scope Creep: Scope creep is something common with every project. This refers to the incremental expansion of the project scope. Most of the time, the client may come back to the service provider during the project execution and add more requirements.

5.6 WORK BREAK DOWN STRUCTURE

- These are incremental and hierarchical decomposition of the project phases. It is a tree like structure and it shown the description of the jobs for completion of the project. These are also called Top-Down diagrams.

- The main jobs are broken down into number of small manageable units depending upon size, time or efforts required for completion. Work Break down may be done in to following stages: 1. Project 2. Subproject 3. Task 4. Subtask 5. Work package.

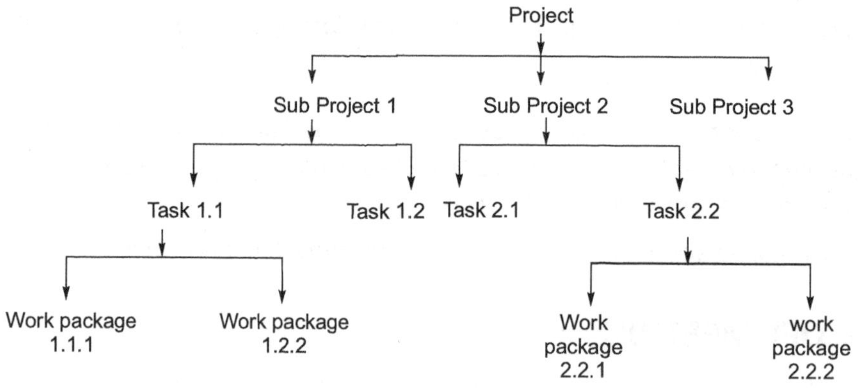

Do you know?: 1. WBS is defined at highest level.
2. In WBS cross functional dependencies is not shown.

5.7 ROLE ASSIGNMENT

It is essential for clarifying and defining roles and responsibilities in projects as everyone on the project needs to understand their role and responsibilities.

The RACI matrix is a responsibility assignment chart that maps out every task, milestone or key decision involved in completing a project and assigns which roles are Responsible for each action item, which personnel are Accountable, and, where appropriate, who needs to be Consulted or Informed. The acronym RACI stands for the four roles that stakeholders might play in any project

5.8 SCHEDULING

This phase overlaps with the planning phase. Time estimates are made for each activity and overall project completion time is calculated. Calculate early and latest, start and finish dates for each activity.

Steps Involved in Planning and Scheduling

1. Determine planning details or work break down structure (WBS)
2. Work break down should be converted into activities and activity inter-relationship should be established.
3. Develop alternative planning strategies and give standardised inputs.
4. Develop resource optimisation considering constraints.
5. Determine activity duration.
6. Prepare project histogram/bar chart/linked bar chart/mile stone chart/network diagram.
7. Make time calculation and find project duration.
8. Review the project progress at regular interval through the process of updating.
9. Monitor the progress regularly.

Schedule Compression Techniques

This are those techniques which are used to reduce project completion time and bring it within the schedule. There are two main techniques as

1. Fast Tracking
2. Crashing

Reason of schedule compression

1. You may be late on schedule
2. You may deliberately want to shorten duration

Causes of delay in Project

1. Unrealistic Schedule
2. Resource unavailability from vendor
3. Some unforeseen incident
4. Customers decision
5 To get away from clashing of launch from competitions project.

Fast Tracking

Here critical path is reduced such that sequential activities run parallelly on partially parallel to each other. Only critical activities are fast track. After fast tracking resheduling is done.

Before fast tracking

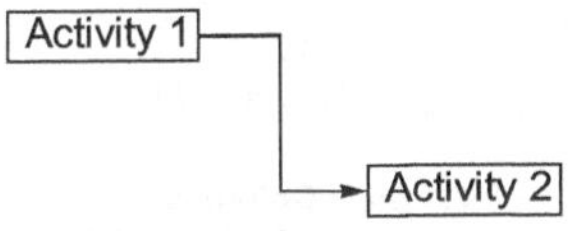

After fast tracking

Crashing

Here extra resources are added to the projects critical activities to compress schedule. Also done only on critical activities, Total cost of project increases. After crashing critical path may change.

Before fast tracking

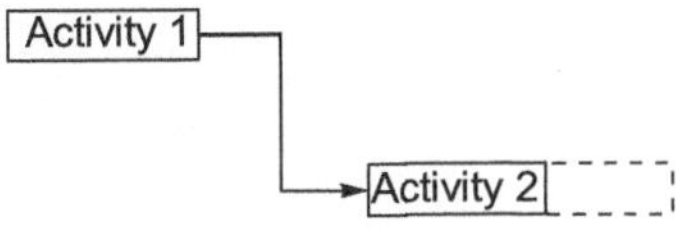

After fast tracking

There is no hard and fast rule for number of levels.

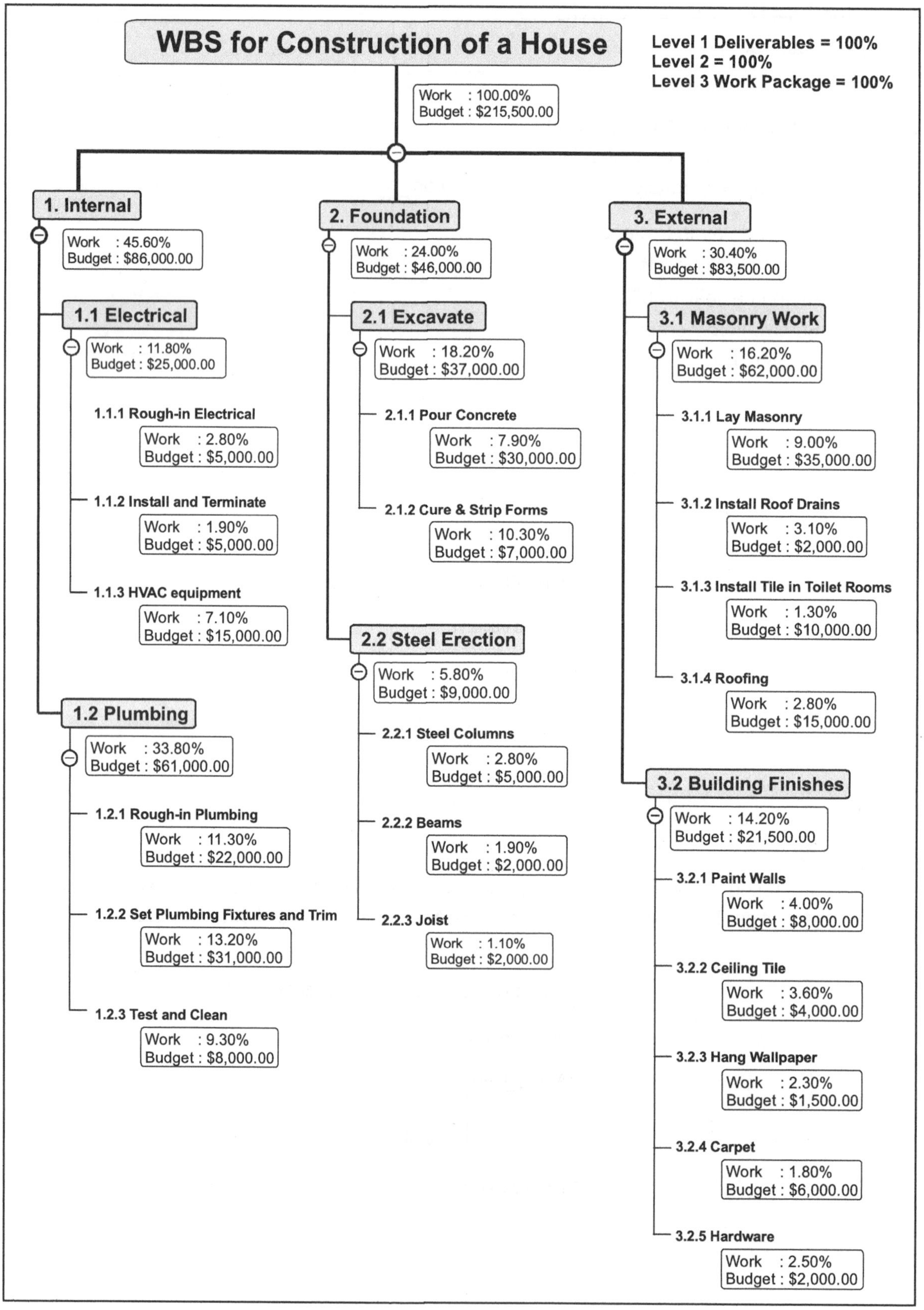

5.9 TECHNIQUES USED FOR SCHEDULING

5.9.1 Bar Charts

Firstly introduced by Henry Gantt around 1900 AD.

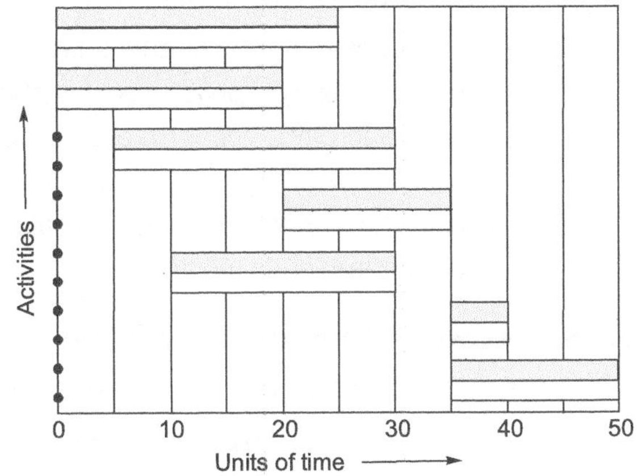

Fig. Bar Chart

Features of bar chart are:

1. It is a pictorial chart.
2. It has two coordinate axes, the horizontal coordinate represents the elapsed time and vertical coordinate represents the job or activity to be performed.
3. The beginning and end of each bar represents starting and finishing time of a particular activity respectively.
4. The length of bar shows the time required for completion.
    - Jobs can be concurrent or can be started one after other. So some bars can run parallel or overlap each other or may run serially.

Limitations of bar chart:

1. *Lack of degree of details*: Only major activities are shown in bar chart and sub-activities can not be separated out. Hence effective control over the activities in big projects can not be achieved.
2. A bar chart does not show progress of work and hence it can not be used as a control device.
3. A bar chart is unable to depict interdependencies of various activities clearly.
4. Bar charts are not useful in the projects where there are uncertainties in determination of estimation of time required for completion of various activities such as in R&D projects.
5. Bar chart can not distinguish between critical and noncritical activities and hence resource smoothening and resource levelling can not be done.

    Bar chart diagrams are useful for only smaller and simpler conventional projects, especially construction and manufacturing projects, in which time estimates can be made with fair degree of certainty.

Do you know?: Higher-level summary activities occurring between milestones in a bar chart are called hammock activities.

5.9.2 Mile-Stone Charts

It is a modification over original Gantt chart. Milestones are key events of main activities represented by bar. Therefore they give idea about completion of sub-activities.

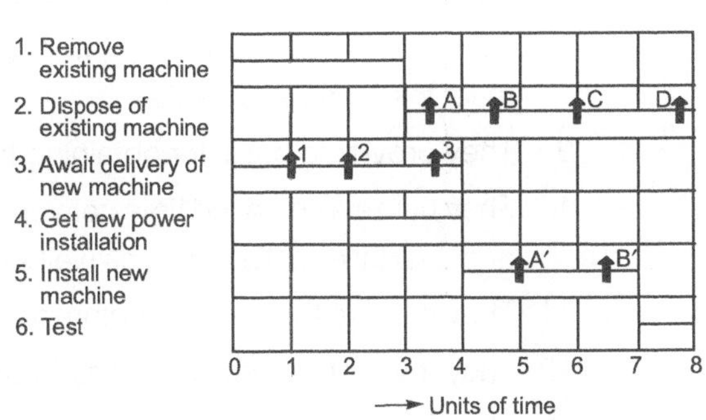

Fig. Mile-stone chart

DO YOU KNOW? Controlling can be better achieved with the help of milestone charts, but still activity interrelationship and accountability of time uncertainty can not be depicted which can be overcome in network technique.

5.9.3 Line of Balance (LOB) Techniques

This is an improvement over mile-stone charts:

- The LOB activity schedule chart shows the graphical plan of work execution in linear or 'S' curve shape. These graph representing the work cycles are referred to as 'cyclographs' or 'cyclograms. The time unit (day or week) in cyclographs are represented along the horizontal axis, while, the vertical axis shows the number of similar work units of the project. The time unit reflected on the horizontal axis can further be divided into calendar months (or weeks) after assessing the working days that are available in each month.

5.9.4 Network Method of Scheduling

- It is an outcome of the improvements in the milestone charts.
- They are called by various names such as PERT, CPM, UNETICS, LESS, TOPS and SCANS.
- However all these have emerged from the two major network systems viz.:
  1. PERT
  2. CPM

5.10 NETWORK DIAGRAM

- It is the flow of diagram consisting of activities and events connected logically and sequentially.
- Network diagram are of two types:
  (i) Activity-on-Arrow Network (A-O-A)
  (ii) Activity-on-Node Network (A-O-N)

Advantages of network method over bar chart and milestone chart:

1. Interrelationships between activities and events of a project are clearly shown.
2. The project can be treated as an integrated whole with all its sub-activities clearly related with each other. It helps in controlling the project.
3. Network method is useful for very complicated projects having large number of activities.
4. It indicates the time required in between two activities in which rescheduling of a project is possible.
5. Time uncertainty is accounted for and so it is also useful for research and development projects.
6. They form the basis for all planning and predicting and help management decide how to use its resources to achieve time and cost goals.
7. They provide visibility and enable management to control "one-of-a-kind" programs.
8. They help management to evaluate alternatives by answering such questions as how time delays will influence project completion, where slack exists between elements, and what elements are crucial to meet the completion date.
9. They provide a basis for obtaining facts for decision-making.
10. They utilize a so called time network analysis as the basic method to determine manpower, material, and capital requirements, as well as to provide a means for checking progress.
11. They provide the basic structure for reporting information.
12. They reveal interdependencies of activities.
13. They facilitate "what if" exercises.
14. They aid in scheduling risk analysis.

5.11 ELEMENTS OF A NETWORK

1. Event

- An event is either start or completion of an activity.
- Events are significant points in a project which act as control points of the project.
- An event is an instant of time and it does not require time or resources.

Following are examples of an event:

1. All parts assembled
2. A budget prepared
3. Construction completed

Following can not be events:

1. Prepare budget
2. Assemble parts
3. Excavate trench

- Events are represented by nodes in a network. It may have any of the following shapes.

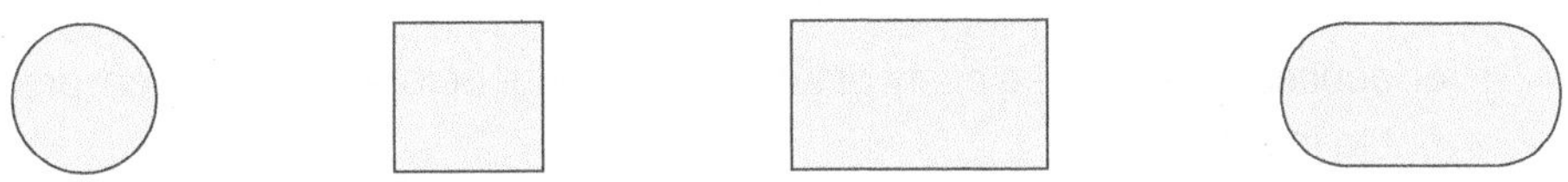

(i) Circular *(ii) Square* *(iii) Rectangular* *(iv) Oval*

Fig. *Different Shapes for Events*

Most commonly adopted shape for events is circular shape.

- **Tail event or the start event:**

  It makes the beginning of an activity.

  If it is the first event of project then known as "initial as start event".

  It has only outgoing arrow.

  E.g: Event 10 is a tail event. Arrows represent job or activity of the project.

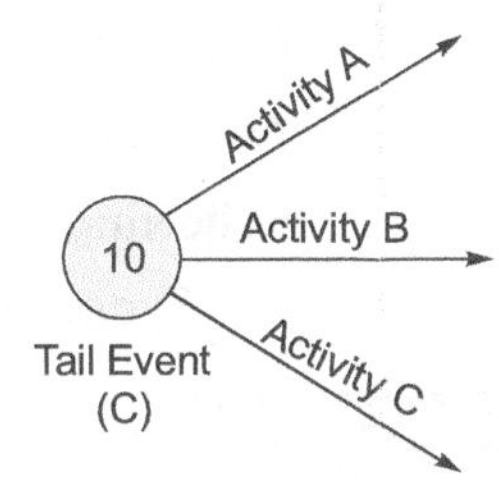

Fig. Tail Event

- **Head event or the final event :**

  The event which marks the completion of an activity is known as "head event".

  If this event represents completion of entire project then it is called "Finish event".

  It has only incoming arrows.

  E.g.: Event 20 is a head event.

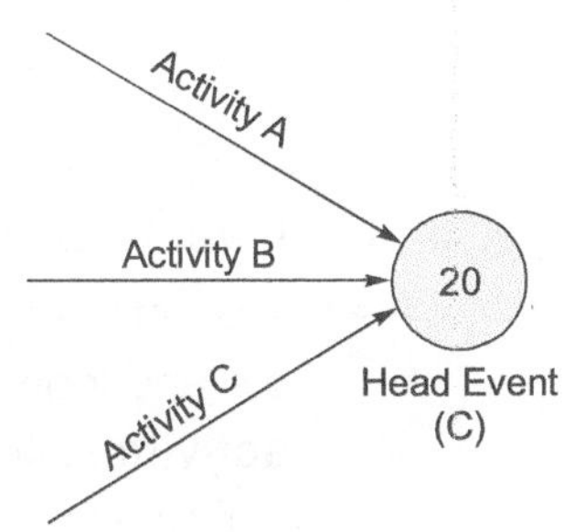

Fig. Head Event

DO YOU KNOW?

(i) When a tail event represents beginning of more than one activity, then the event is said to occur when the first activity starts from it.

(ii) Similarly, when a head event occurs at the end of more than one activity, the event is said to have occurred only when all the activities leading to it, are completed.

- **Dual role events:** All events except the first and the last event of a project are dual role events. They have both incoming and outgoing arrows.

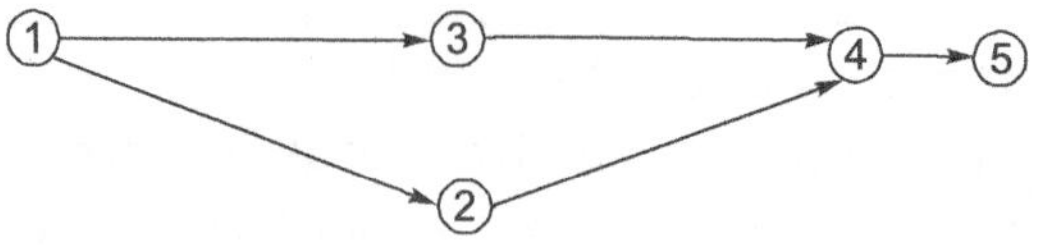

Fig. *Dual Role Events*

E.g.: Events 2, 3 and 4, are dual role events.

- **Successor events:** The event or events that follow another event are called successor events to that event.

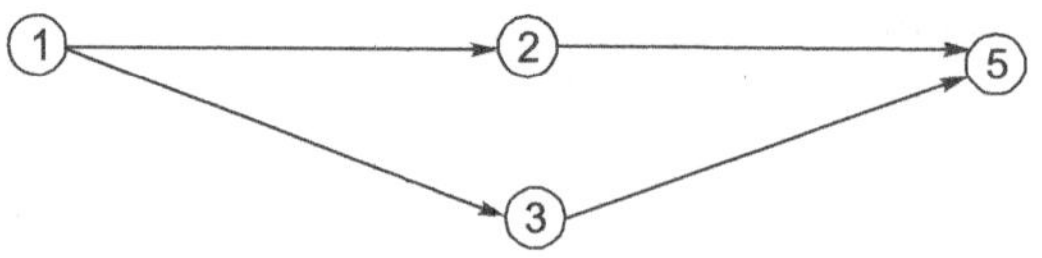

Fig. *Successor Events*

E.g.: Event 2 and 3 are successor events of event 1.

- **Predecessor events:** The event or events that occur before another event are called predecessor event to that event.

  In above figure, events 2, 3 are predecessor to event 5.

Do you know?: It should be noted there can be only one tail event and one head event in a project.

2. Activity

Activity is actual performance of a job. It requires time and resources for its completion.

Following are examples of an activity:

1. Excavate trench
2. Mix concrete
3. Prepare budget

DO YOU KNOW?

In A-O-A system (Activity On Arrow network system), activity is represented by arrows between events while in A-O-N (Activity On Node system), activities are represented by nodes. In A-O-N system, events have no places.

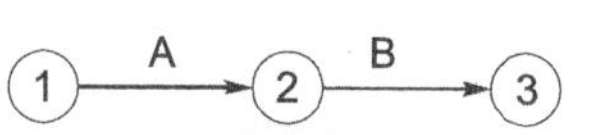

A-O-A

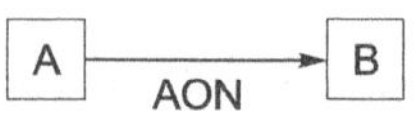

Here *A* & *B* activities are represented in two different systems.

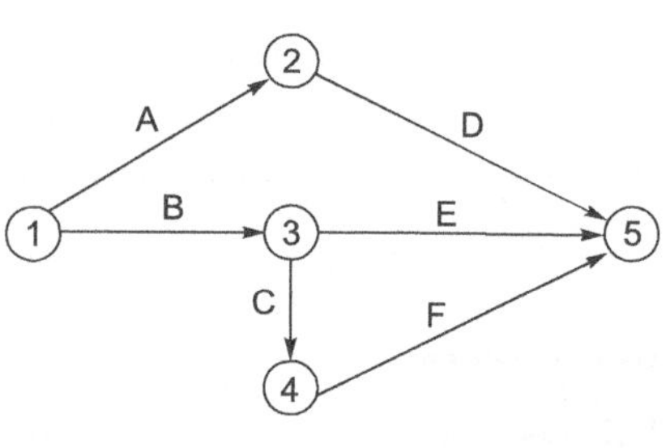

- The activities which can be performed simultaneously are and independent of each other called as **parallel activities**. In figure, activities *A* & *B* are parallel activities.
- Activity or activities that can be performed after performance of other activity are known as **successor activities** to that activity. Activity *F* is successor activity to activity *C* in figure.
- Similarly activities that are required to be performed before another activity can begin are called predecessor activities to that activity. Activity (*A*) is predecessor activity to activity *D*.

3. Dummy

- A dummy is a type of operation which neither requires time nor any resource, but it denotes dependency among the activities.

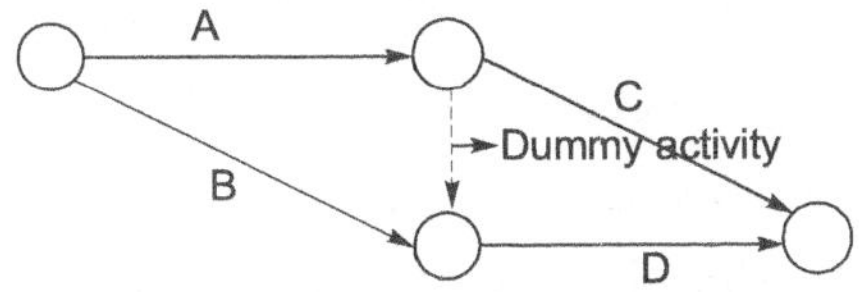

Fig. *Dummy Activity*

- It is represented by dashed arrow.
  In the figure shown below, a dummy activity is shown.
- Dummy is used to serve following purposes:
  1. ***Grammatical purpose*** : To prevent two arrows having common beginning and common end.

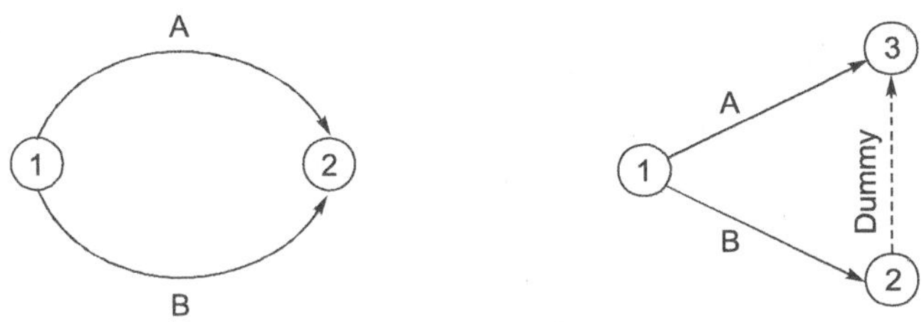

(a) Ambiguous Representation *(b) Grammatically Clean Representation*

Fig. *Geometrical Purpose*

  2. ***Logical purpose*** : To show relationship with other activities.
     Here dummy is required to show that activity *D* can start after completion activities of *A* & *B* both.
     Unnecessary dummies should be avoided.

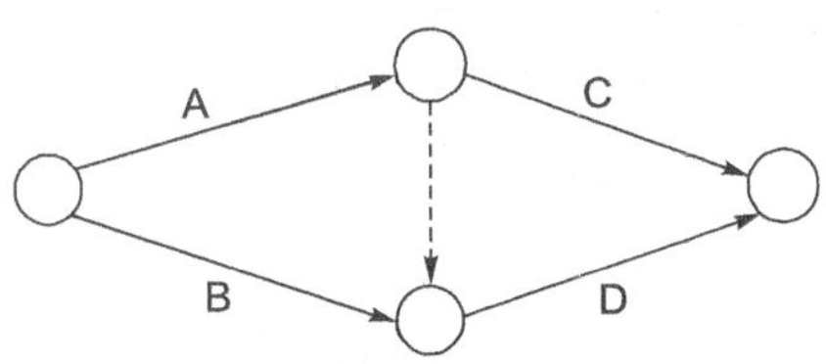

Fig. *Logical Purpose*

> **DO YOU KNOW?**
> - Dummies are used to show predecessor relation but if that relation is already established in the network, then that dummy is redundant and has to be removed.
> - *If dummy is only incoming/outgoing arrow to/from a node then it can be remove provided there is no logical or grammatical error.*

5.12 RULES OF A NETWORK

1. There can be only one initial and one final event.
2. An event can not occur unless all preceding activities are completed.
3. An event can not occur twice.
4. Number of arrows should be equal to number of activities.
5. Time should always flow from left to right.

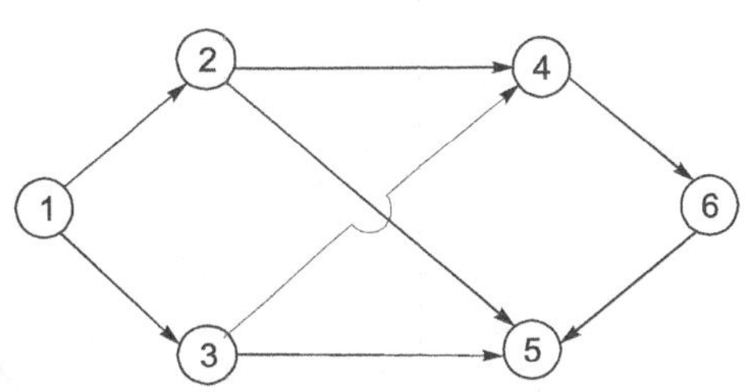

Fig. *Crossing Activities*

6. Length of arrow does not show any magnitude. Straight arrows should be taken as far as possible.
7. Arrows should normally not cross each other. If it is necessary to cross, one should be bridged over the other.
8. No activity can start until its tail event has occurred.

Fulkersons's Rule for Numbering the Events:

1. The single initial event is numbered as 0, 1, 10 etc.
2. All arrows emerging out of the initial event are neglected. Doing so, the created one or more new initial events are numbered as 2,3,4 or 20, 30, 40 etc.
3. Step - 2 is repeated unless all events are numbered.

Errors in Network:

1. ***Looping error*** : Loops should not be formed.

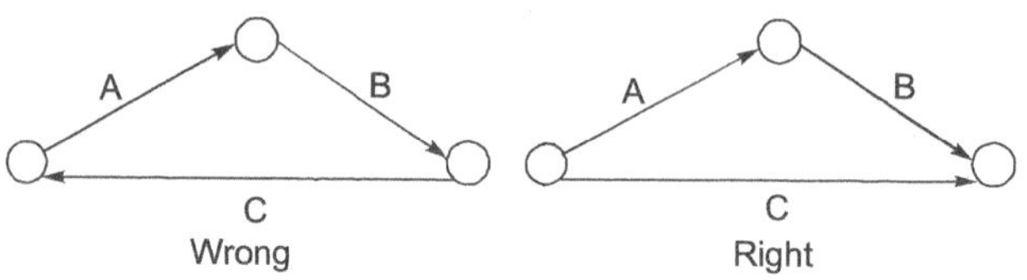

Fig. *Looping Error*

2. ***Dangling error*** : Project is complete only when all its activities are complete but the duration of activity '*R*' has no effect on the project time as shown in figure.

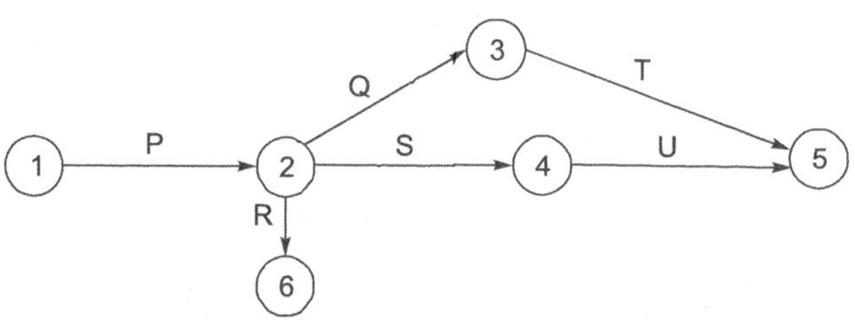

Fig. *Dangling Error*

To avoid dangling error, the network must be examined in such a manner that all events except initial and final events must have at least one activity entering and one activity leaving them.

3. ***Wagon wheel error :*** As shown in figure below, each of the activities *P*, *Q* and *R* cannot start until all the three activities *A*, *B* and *C* are completed. But in reality, this may not be the situation. There is no error visible in the construction of diagram but logical error has crept into it. If all these activities *A*, *B* and *C* are not precedent to *P*, *Q* and *R*.

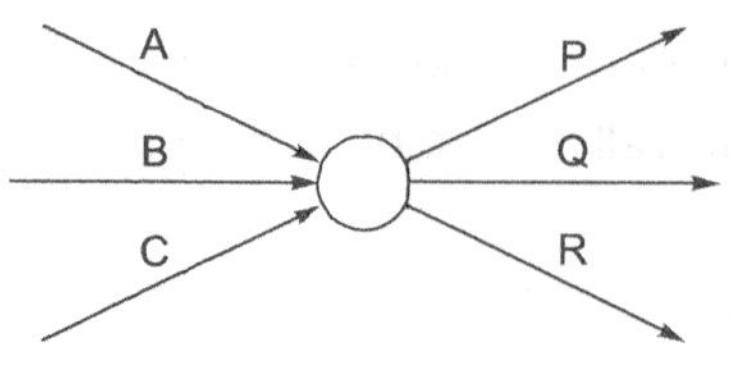

Fig. *Wagon Wheel Error*

5.13 NETWORK SCHEDULING TECHNIQUES

1. Program Evaluation and Review Technique (PERT)
2. Arrow Network Diagram (AND) or Critical Path Method (CPM)
3. Precedence Diagram Method (PDM)
4. Graphical Evaluation and Review Techniques (GERT)

5.13.1 Programme Evaluation and Review Technique (PERT)

- PERT stands for "Project/Programme Evaluation and Review Technique".
- PERT involves uncertainty into the project completion time.
- It is a numerical technique used in the projects in which time can not be estimated accurately such as research and development projects. It is an event **oriented network**. Cost is assumed to be directly proportional to time.

Three time estimates are made in PERT:

1. **Optimistic time (t_o):** This is the minimum possible time in which an activity can be completed under the most ideal conditions.
2. **Pessimistic time (t_p):** This is the maximum time required to complete an activity under the worst possible conditions.
3. **Most likely time (t_m) :** This is the time required to complete an activity under normal working conditions. Its value lies between t_o and t_P. It is near to the expected time.

DO YOU KNOW?

The most likely time (t_m) is based on experience and judgement being based on the time required if the activity is repeated a number of times under essentially the same conditions. This time signifies the most frequently occurring time. It reflects a situation "things are as usual, nothing exciting".

Mean Time or Expected Time or Average Time

- In PERT each activity is assumed to follow β-distribution curve of time.
- This is calculated from β-distribution curve of time at which probability of activity is just 50%. Time taken by various activities follow β-distribution.

  Hence value of expected time is calculated by weighted average as,

$$t_e = \frac{t_o + 4t_m + t_p}{6}$$

Standard Deviation of an Activity (σ)

- This is the measurement of uncertainty, which is approximately one sixth of time range i.e.,

$$\sigma = \frac{t_P - t_o}{6}$$

- It can be seen above that 'σ' is affected by relative distance from the most optimistic estimate to the most pessimistic estimate.
- Therefore, wide range in time estimates represents greater uncertainly.

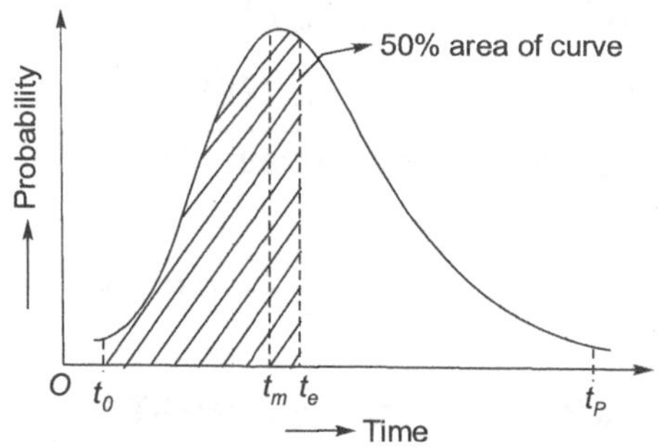

Fig. β-*curve with skewness to the left*

> **DO YOU KNOW?**
> In a limiting case, certainty of an activity duration occurs only when the three time estimates coincide, so that the standard deviation and the variance both vanish. Consequently the activity duration becomes certain which is the case of CPM. Hence, a PERT is a general case whereas CPM is the particular case of PERT.

Variance of an Activity (σ^2)

Square of standard deviation is variance of an activity. It is to be noted that higher the uncertainty about a process, greater is the standard deviation and hence greater is the variance of a project.

> **DO YOU KNOW?**
> A beta distribution is the one which is not symmetrical about its apex. Figure below shows two beta distribution curves, one having skew to the left (Beta distribution for optimistic estimator) and other having skew to the right (Beta distribution curve for the pessimistic estimator).

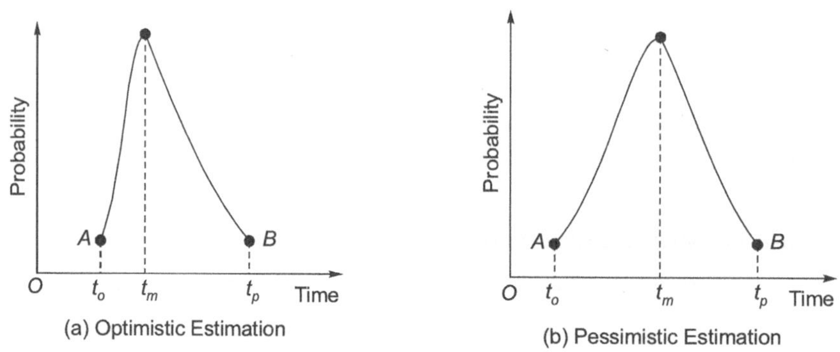

Fig. *Two different forms of β-distribution curves*

Central Limit Theorem

The theorem states that a project consists of a large number of activities, where each activity has its own mean time (t_e), standard deviation (σ), variance (σ^2) and also its own β-distribution curve. The distribution of time for the project as a whole will approximately be a normal distribution, i.e. mean time or expected time of a project is

$t_e = t_{e_1} + t_{e2} + t_{e3} + ...$ along critical path and the variance is,

$$\sigma^2 = \sigma_1^2 + \sigma_2^2 + \sigma_3^2 + ...$$

along critical path.

Hence standard deviation of the project as a whole

$$\sigma = \sqrt{\sigma_1^2 + \sigma_2^2 + \sigma_3^2 + ...}$$

along critical path.

Critical path: The time wise longest path is the critical path. In this path, any type of delay in any event will cause delay in the project. These are shown by double lines or dark lines in a network.

1-2-3-4 is the critical path of following network.

Time computation of Events:

1. **Earliest Expected Occurrence Time (EOT):** The time at which an event is expected to occur earliest.
   - The time at which an event can be expected to occur earliest.
   - An event occurs when all the activities leading to it are completed.
   - It is generally denoted by T_E. It is calculated by forward path.

$$T_E^j = T_E^i + t_{ij} \qquad \text{...(when there is only one path)}$$

$$T_E^j = (T_E^i + t_{ij})_{max} \qquad \text{...(when there is more than one path)}$$

Here,
T_E^j = EOT of event j
T_E^i = EOT of event i
t_i^j = Expected time of activity i-j

2. **Latest Allowable Occurrence Time (LOT):** The latest allowable time at which an event must occur to keep the project on schedule.
   It is generally denoted by T_L. This is calculated through backward path.

$$T_L^i = T_L^j - t_{ij} \qquad \text{...(when there is only one path)}$$

$$T_L^i = (T_L^j - t_{ij})_{min} \qquad \text{...when there is more than one path)}$$

where
T_L^j = LOT of event j
T_L^i = LOT of event i
t_{ij} = expected time of activity i-j

NOTE : The latest allowable occurrence time of the finish event is equal to the schedule completion time of the project.

- **Slack:** Slack is defined as the difference between latest allowable time (T_L) and earliest expected time (T_E) of an event.
  Slack for any event j, $S_j = T_L^j - T_E^j$
  Slack for any event i, $S_i = T_L^i - T_E^i$
- Slack may be positive, zero or negative.
- When "Slack is greater than zero". It indicates project is ahead of schedule and availability of excess resources. Such events are sub critical.
- If slack is zero, it indicates work is on schedule and events are critical. Resources are just adequate.
- If slack is negative, it indicates work is behind schedule and may cause delay in project completion. Events are super critical. Extra resources are required.
- The path having minimum or zero slack value is the "critical path" which is also time wise longest path.

DO YOU KNOW?

If a project has more than one critical path then severity is calculated by calculating standard deviation of the project from each path. (Having lesser value of s have more deterministic approach hence critical path will be having more value of σ).

Probability of a project to be completed in schedule time:

- Probability of a project completion in expected time (T_e) is just 50% and probability of completion increases or decreases with an increase or decrease of scheduled time respectively.
- As the probability distribution curve for the entire project is normal distribution curve.
- Probability can be calculated by using probability factor (Z) which is defined as,

Value of Z	Probability (%)
0	50.00
+1	84.13
+2	97.72
+3	99.87
–1	15.87
–2	2.28
–3	0.13

$$Z = \frac{T_s - T_e}{\sigma}$$

It should be noted that:

1. At $Z = 0$, $T_s = T_e$ and probability = 50%
   $Z < 0$, $T_s < T_e$ probability < 50% and vice versa.
2. For intermediate value of Z, interpolation is required.
3. The time and probability distribution for project as a whole is a normal distribution.
4. Area under the curve is equal to unity.
5. 68% of the area lies within '$t_e - \sigma$' and '$t_e + \sigma$' range. This is called probable range.
6. Area bounded by $t_e + 2\sigma$ and $t_e - 2\sigma$ is about 95%.
7. Area bounded by $t_e - 3\sigma$ and $t_e + 3\sigma$ is about 99.7% which is approximately equal to area bounded by t_o and t_p, which is maximum probable range.

$$\sigma = \frac{t_p - t_o}{6}$$

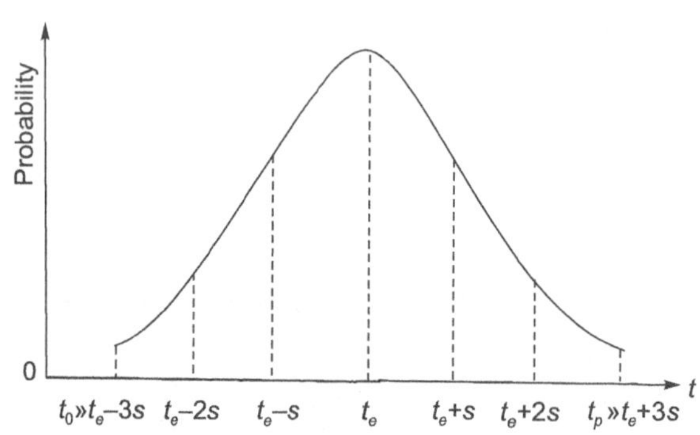

Fig. *Probability Distribution Curve*

5.13.2 Critical Path Method (CPM)

This is based on deterministic approach in which only one time estimate is made for activity completion. Network diagram in CPM is activity oriented.

- It is activity oriented network.
- It is used for repetitive type of work and has deterministic approach.

Event Times in CPM

1. Earliest occurrence time (T_E): Time at which an event may occur as early as possible.
2. Latest allowable time (T_L): Time at which event may occur as late as possible without delaying the overall project completion time.

   These are similar to PERT and are calculated in the same fashion.

Activity Times in CPM

1. **Earliest start time (EST) :** It is the earliest possible time at which an activity can be started.

   For an activity *i* – *j*, earliest event time of event *i*, i.e. T_E^i is EST of activity *i* – *j*.

2. **Earliest finish time (EFT) :** It is the earliest possible time by which an activity can be completed. For an activity *i* – *j*

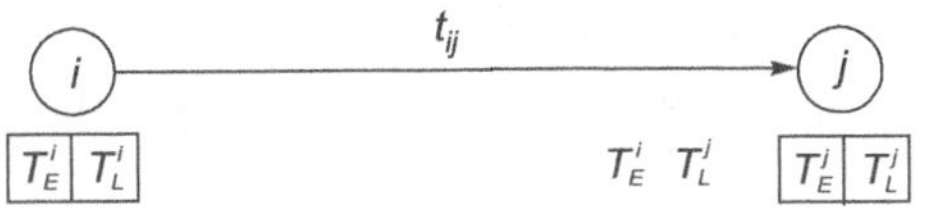

Fig. *EST and EFT*

$$EFT = EST + t_{ij} = T_E^i + t_{ij}$$

$$t_{ij} = \text{Activity duration}$$

3. **Latest start time (LST) :** This is the latest possible time at which an activity can be started without delaying the overall project.

   ∴ LST = LFT – Activity duration

   ∴ $LST = T_L^j - t_{ij}$

   LFT = Latest finish time of activity *i* – *j* = T_L^j

4. **Latest finish time (LFT) :** This is the latest time by which an operation or activity must be completed without delaying the project.

   For an activity *i* – *j*, latest allowable time of head event *j*, i.e. T_{Lj} is LFT of activity *i* – *j*.

Do you know?: LST of an activity is to be calculated on the basis of latest occurrence time of its head event and not on the basis of latest occurrence time of its tail event.

Float

- It is associated with activity times.
- It is analogous to slack of events in PERT.
- It is the range within which start or finish time of an activity may fluctuate without affecting the project completion time.
- Floats are of following types:

  1. **Total float:** The time span by which starting or finishing of an activity can be delayed without delaying the completion of the project.

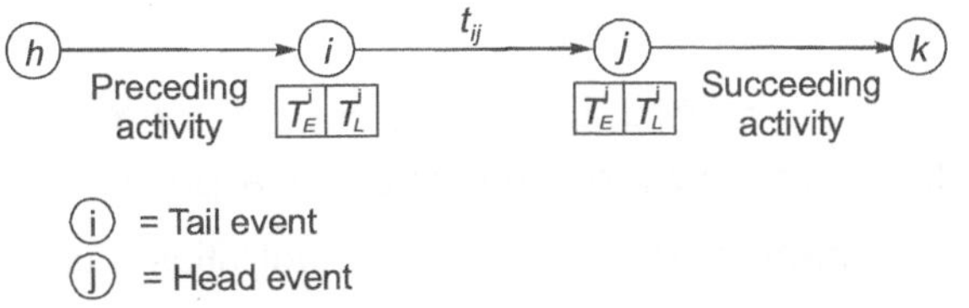

Fig. *Total float*

It is the maximum available time in excess of the activity completion time.

Total float is given by F_T

$$F_T = (T_L^j - T_E^i) - t_{ij}$$

Total float is given by F_T

$$F_T = (T_L^j) - (T_E^j + t_{ij})$$
$$F_T = (T_L^j) - (T_E^j + t_{ij})$$
$$F_T = \text{LFT} - \text{EFT}$$

or
$$F_T = (T_L^j - t_{ij}) - T_E^i$$
$$F_T = \text{LST} - \text{EST}$$

- Total float of an activity affects total float of succeeding as well as preceding activities.

Do you know?: Total float of an activity constraints the finishing of preceding activity and starting of succeeding activity.

2. **Free float (F_F) :** The delay which can be made without delaying succeeding activities. It affects only preceding activities. It is denoted by F_F. It is assumed that all activities start as early as possible.

   Free float is given by

$$F_F = (T_E^j - T_E^i) - t_{ij}$$

$\Rightarrow$
$$F_F = F_T - S_j$$

where S_j is head event slack.

DO YOU KNOW?

In free float, preceding activity is not allowed to occur at its latest time and hence total float of preceding activity is affected. However the succeeding activity can start at its earliest start time and hence its total float is not affected.

3. *Independent Float (F_{ID}): It is the minimum excess available time which exists without affecting any of* succeeding or preceding activities. It is denoted by F_{ID}.

   - It is the excess of minimum available time over the activity duration.

$$F_{ID} = (T_E^j - T_L^i) - t_{ij}$$
$$F_{ID} = F_F - S_i$$

   where S_i is tail event slack.

4. **Interfering float (F_{INT}):** It is similar to head event slack.

$$F_{INT} = S_j = F_T - F_F$$

Critical Path

In CPM analysis, the path along which total floats are zero or minimum is called as critical path. All activities on this path are critical. There can be more than one critical paths.

Subcritical Path: It is the path joining all subcritical activities. For a subcritical activity total float is greater than zero i.e.,

$$F_T > 0$$

Supercritical Path: It is the path joining all super critical activities. For a supercritical activities total float is less than zero i.e.,

$$F_T < 0.$$

Comparison between PERT and CPM

PERT	CPM
1. It is event oriented.	1. It is an activity oriented network.
2. It has probabilistic approach. The probability distribution is of the type of β distribution	2. It has deterministic approach. Probability value approaches to one here.
3. Three types of times are estimated on the basis of which an expected time t_e is derived.	3. Only one time is calculated i.e. activity duration 't'.
4. Cost is directly proportional to time. Hence efforts are made to minimize the time so as to result in the minimum cost.	4. Time and cost are related by the following curve given. From this curve optimum time is derived which results in the minimum cost.
5. It is suitable for newer type of projects which have not been performed in the past and no exact assessment of time and cost is available. **Examples:** Research work, launching of space aircraft, development of missile programme etc.	5. It is suitable for repetitive type of work where time and cost can be evaluated with fair degree of accuracy.
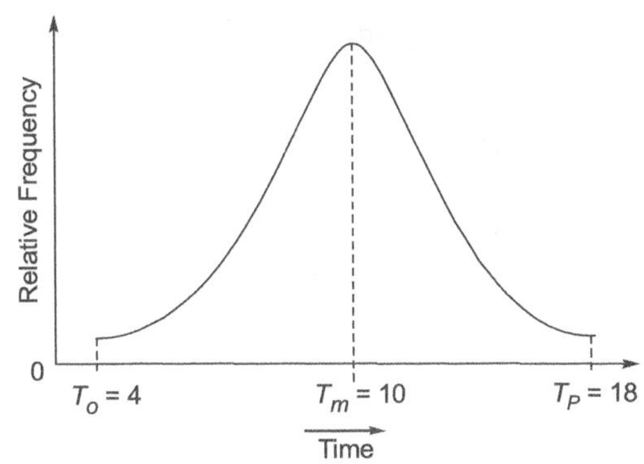	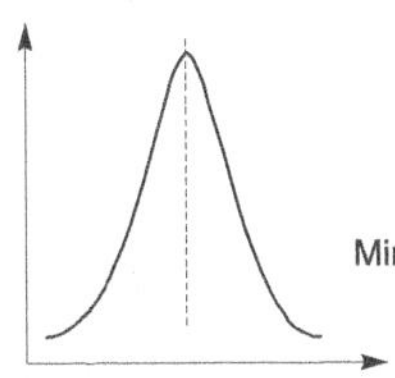 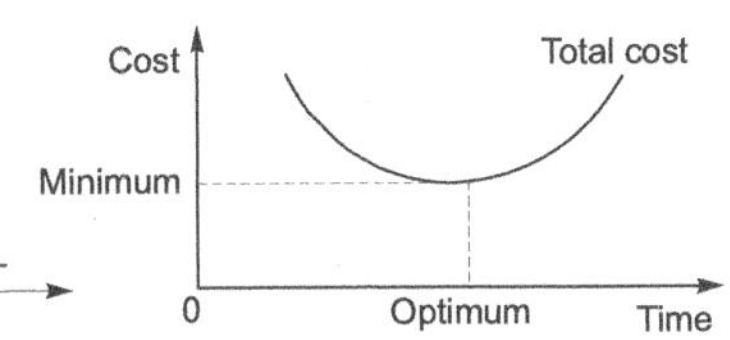 *Time-cost Relationship* **Examples:** • Construction work, • Maintenance work, • Civil engineering projects.

5.13.3 CPM Systems

Mainly two systems are used in CPM analysis:

1. **A-O-A system** (Activity on arrow system)

   An activity is graphically represented by an arrow.

   The tail end and head end of arrow represents start and finish of an activity respectively.

Do you know?: Dummies have been used to represent the constraint and consequent interdependence. But number of dummies must be minimum for an efficient network.

2. **A-O-N System** (Activity on node system or precedence diagram). Activity is represented by a circle or a node. Events have no places. Arrows are used only to show the dependency relationship between activity nodes.

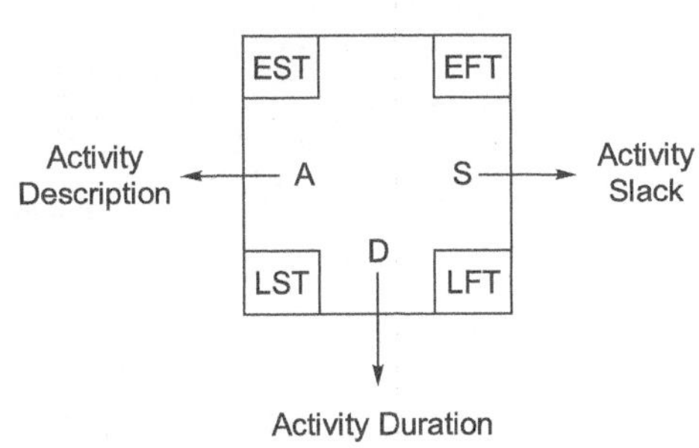

Fig. Debut Activity

- When two or more activities start parallely then an activity called DEBUT (D_O) is provided at the beginning. Like wise a finish activity (F_O) is provided at the end when more than one activities finish parallely. Activity D_O & F_O have zero duration.

Advantages of A-O-N System over A-O-A System

1. A-O-N system eliminates the use of dummy activities.
2. It is more helpful for projects having more overlapping activities.
3. It is self sufficient and self-explanatory. All activity times (EST, EFT, LST, LFT) are represented on the diagram.
4. Revision and modifications are easier.
5. Pre-operations and post-operations of activities under consideration are distinctly visible.

Examples:

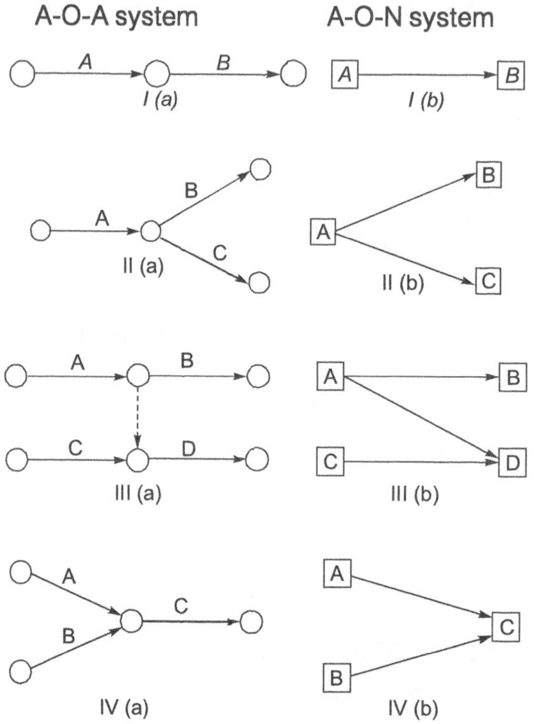

EXAMPLE 5.1

Consider the network as shown below with expected time of completion of each activity (t_E) (in days) is shown on arrows. Calculate the earliest expected time of the network for the completion of work.

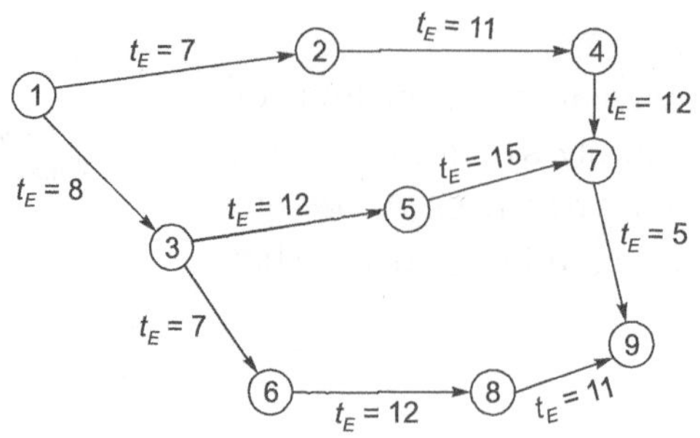

Solution:

Here event 7 is approached by two paths viz. path along 1 → 2 → 4 → 7 and 1 → 3 → 5 → 7.

For path 1 → 2 → 4 → 7,

$$t_E \text{ for event } 7 = 7 + 11 + 12 = 30 \text{ days}$$

For path 1 → 3 → 5 → 7,

$$t_E \text{ for event } 7 = 8 + 12 + 15 = 35 \text{ days}$$

Thus event 7 cannot be considered to occur until all activities of path 1 → 3 → 5 → 7 have completed i.e., event 7 cannot occur prior to 35 days. Thus earliest expected time for event 7 is 35 days.

In this network, event 9 is the terminal event which is preceded by path 7 → 9 and 3 → 6 → 8 → 9.

For path 7 → 9,

$$T_E \text{ for event } 9 = 35 + 5 = 40 \text{ days}$$

For path 3 → 6 → 8 → 9

$$T_E \text{ for event } 9 = 8 + 7 + 12 + 11 = 38 \text{ days}$$

Thus earliest expected time of event 9

= Earliest expected time for completion of work

= 40 days (larger of two values viz. 40 days and 38 days)

EXAMPLE 5.2

In the network as shown, the three times viz. optimistic time, most likely time and the pessimistic time (in days) are shown on the arrows. Compute the earliest expected times and the latest allowable occurrence times of various events.

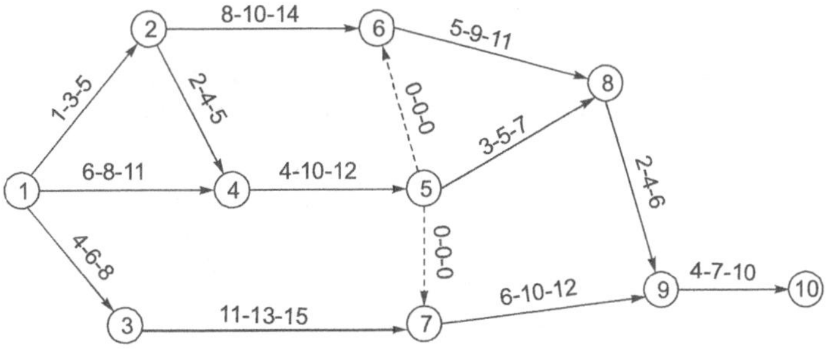

Solution:

The expected time of an activity is given by,

$$t_E^{ij} = \frac{t_0^{ij} + 4t_L^{ij} + t_P^{ij}}{6}$$

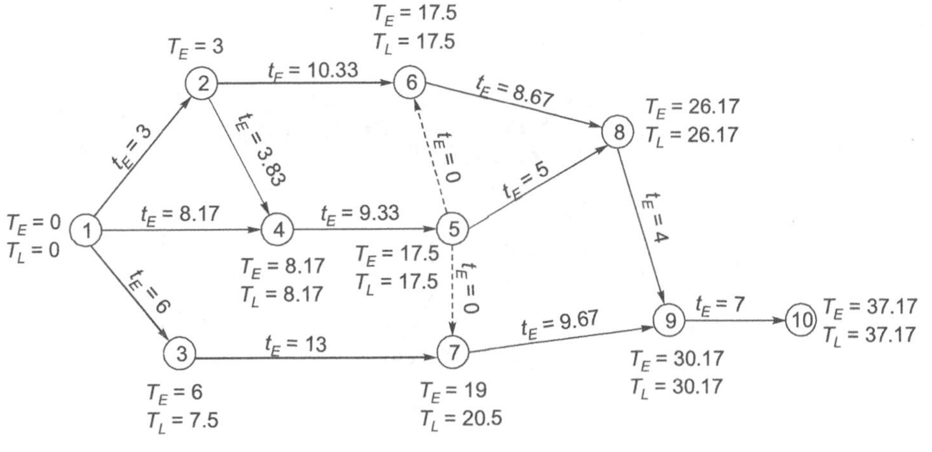

Activity i-j		t_o^{ij}	t_L^{ij}	t_P^{ij}	t_E^{ij}	T_E^j	T_L^j
Successor event j	Predecessor event i						
10	9	4	7	10	7	37.17	37.17
9	8	2	4	6	4	30.17	30.17
9	7	6	10	12	9.67	30.17	30.17
8	6	5	9	11	8.67	26.17	26.17
8	5	3	5	7	5	26.17	26.17
7	5	0	0	0	0	19	20.5
7	3	11	13	15	13	19	20.5
6	5	0	0	0	0	17.5	17.5
6	2	8	10	14	10.33	17.5	17.5
5	4	4	10	12	9.33	17.5	17.5
4	2	2	4	5	3.83	8.17	8.17
4	1	6	8	11	8.17	8.17	8.17
3	1	4	6	8	6	6	7.5
2	1	1	3	5	3	3	4.34

EXAMPLE 5.3

A network is shown below with activity times (in days) marked on the arrows. Compute the various floats.

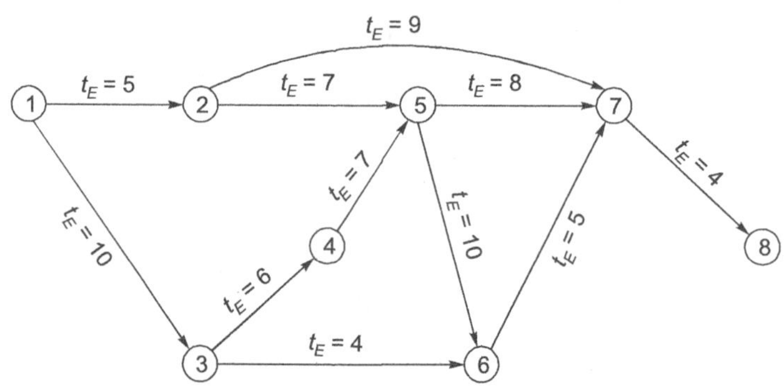

Solution:

The earliest start time (EST) and latest finish time (LFT) of each event is computed and is shown in the network.

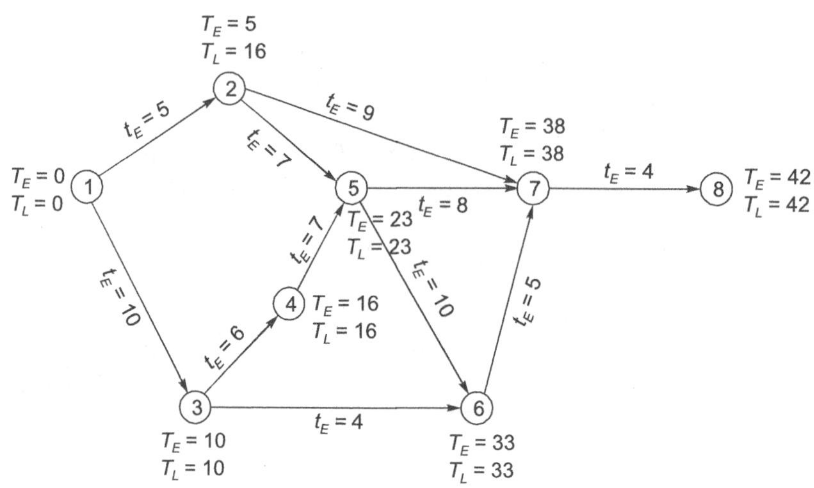

Activitiy (*i–j*)	Duration (days) t^{ij}	Earliest Time		Latest Time		Total Float (F_T)	Free Float (F_F)	Independent Float (F_{ID})
		Start (EST)	Finish (LFT)	Start (LST)	Finish (LFT)			
1-2	5	0	5	11	16	11	0	0
1-3	10	0	10	0	10	0	0	0
2-5	7	5	12	16	23	11	11	0
2-7	9	5	14	29	38	24	24	13
3-4	6	10	16	10	16	0	0	0
3-6	4	10	14	29	33	19	19	19
4-5	7	16	23	16	23	0	0	0
5-6	10	23	33	23	33	0	0	0
5-7	8	23	31	30	38	7	7	7
6-7	5	33	38	33	38	0	0	0
7-8	4	38	42	38	42	0	0	0

Critical path will have total float as zero for each activity.

∴ Critical path is $1 \rightarrow 3 \rightarrow 4 \rightarrow 5 \rightarrow 6 \rightarrow 7 \rightarrow 8$.

NOTE

- Slack condition gives only necessary condition while float concept provide both necessary and sufficient condition.
- *As in above example, event (5) and (6) having zero slack but activity 5–7 is not critical.*

EXAMPLE 5.4

Find out the project completion time and critical path for the network shown below. The activity durations are in days.

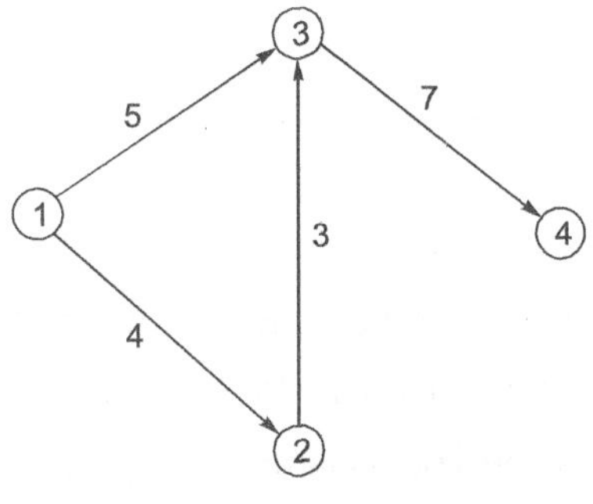

Solution:

The network can be redrawn as shown below:

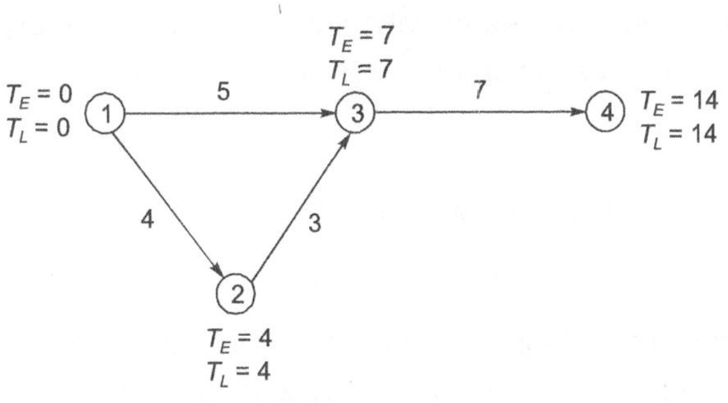

Activity (i-j)	Duration (t^{ij}, days)	Earliest Time		Latest Time		Total float (F_T)
		Start (EST)	Finish (EFT)	Start (LST)	Finish (LFT)	
1-2	4	0	4	0	4	0
1-3	5	0	5	2	7	2
2-3	3	4	7	4	7	0
3-4	7	7	14	7	14	0

$\therefore$ Project completion time = 14 days.

Critical is the one which is having zero total floats i.e., path $1 \rightarrow 2 \rightarrow 3 \rightarrow 4$ is critical.

5.14 PROJECT COST CONTROL

5.14.1 Time-Cost Trade-off

- Time and cost are the two most important resources that a project manager deals with. Both these resources have constraints and the job of a project manager is to have a judicious balance between the two. The judicious balance between time and cost is called Time-Cost Trade-off and it can be achieved by studying the availability of these two resources and the demand for these two resources for the given project.
- Every project is unique and the same time-cost trade-off rule will not be applicable to all projects uniformaly. For example,
  - (i) The duration for a project as obtained from network calculations might be longer than the desired duration. Social, environmental or political constraints may necessitate early completion of the project even at a higher cost.
  - (ii) When cost is the only constraint and there is no necessity that the project should be hurried through, the project time can be extended till the total-cost becomes the minimum.

5.14.2 Cost Model Analysis

- In CPM, time is related to cost and the objective is to develop an optimum time-cost relationship.
- The overall project duration can be reduced by reducing the duration of only the critical activities in the project network. The durations of such activities may be reduced in two ways.
  - (a) by deploying more resources for the early completion of such activities.
  - (b) by relaxing the technical specifications for such activities.
- In whole of CPM Cost Model, we will be assuming that project duration is reduced by deploying more resources on critical activities but not relaxing specifications.
- In CPM, there are two factors, time and cost estimates for each activity: 'normal estimate' and crash estimate'. In the normal estimate, the emphasis is on cost with time being associated with minimum cost. The 'crash' estimate involves the absolute minimum time required for the job and the cost necessary to achieve it. Here the emphasis is on 'time'.

Project Cost

The components of the total cost are shown in figure.

(a) Indirect Project Cost

- Indirect costs on a project are those expenditures which cannot be apportioned or clearly allocated to the individual activities of a project, but are assessed as a whole. The indirect cost includes the expenditure related to administrative and establishment charges, overhead, supervision, expenditure on a central store organisation, loss of revenue, loss of profit, penalty etc.

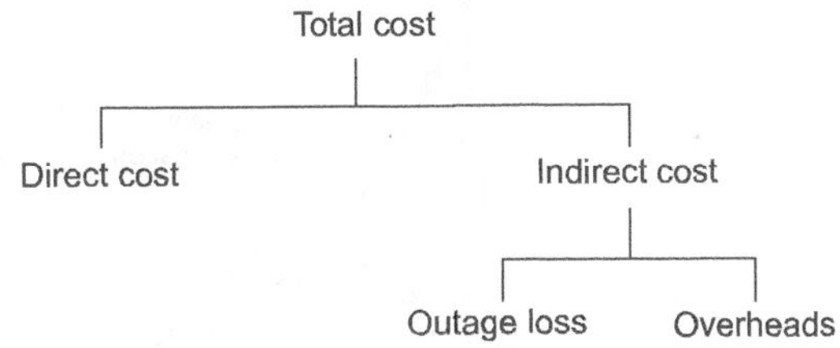

Fig. *Project cost*

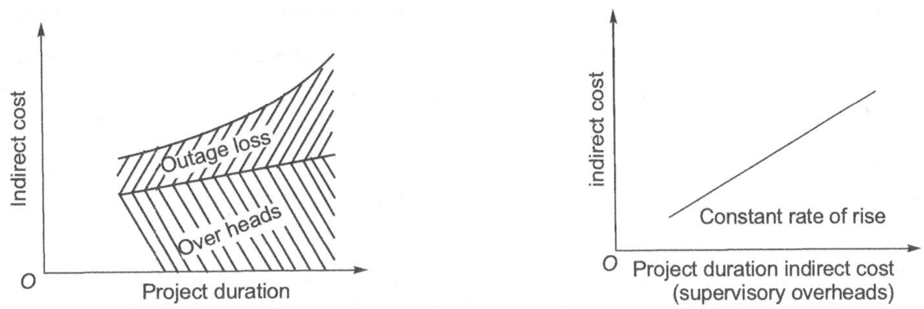

Fig. *Indirect Project Cost*

- Indirect cost rises with increased duration, considering only overhead and supervision. It is represented by a straight line, with a slope equal to daily overhead.
- But when there is a loss in profits, due to inability to meet demand or due to some penalty due to delay, a corresponding cost increase must be added to the cost of overheads, producing the curve. Such a loss is called the outage loss.
- The total indirect cost curve will thus be curved however for purpose of computation it is assumed linear.

(b) Direct Project Cost

- It is the cost which is directly dependent on the amount of resources involved for completion of activities. It includes labour, materials, plants and machining etc.
- To get the same work done in less time, we have to increase amount of labour, equipment and time saving material that too at extra charges which simply means increase in direct cost.
- The project has the highest direct cost corresponding to the crash duration, and has normal cost corresponding to the normal duration.

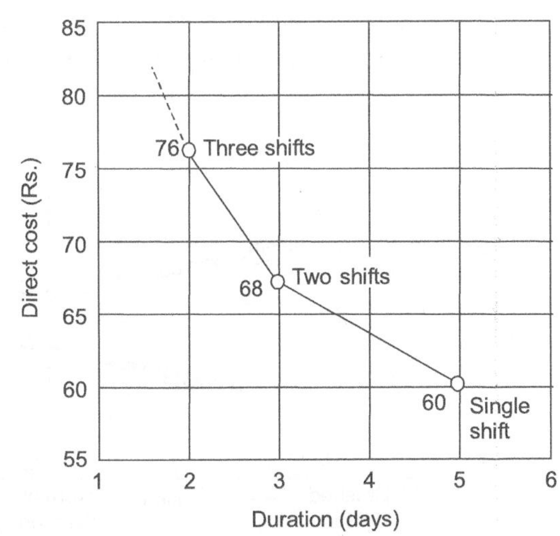

- **Normal time (t_n)** : Normal time is the standard time that an estimator would usually allow for an activity.
- **Crash time (t_c)** : Crash time is the minimum possible time in which an activity can be completed, by employing extra resources. Crash time is that time, beyond which the activity cannot be shortened by any amount of increase in the resources.
- **Normal cost (C_n)** : This is the direct cost required to complete the activity in normal time duration.

- **Crash cost (C_c):** This is the direct cost corresponding to the completion of the activity within crash time.

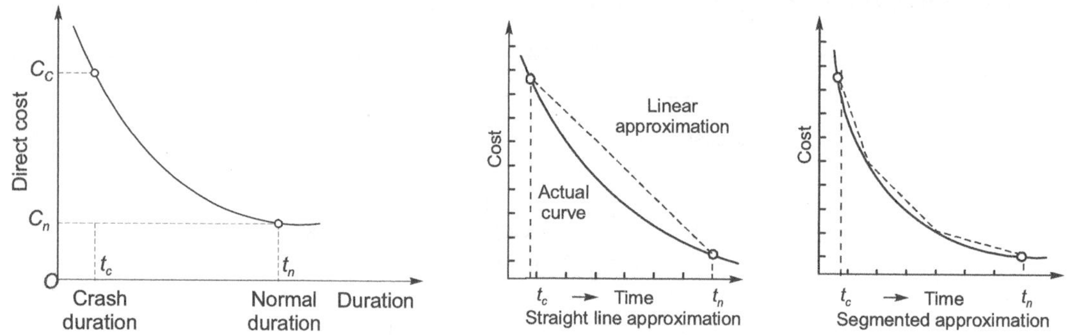

Generalized direct cost-time curve *Direct cost curve approximation*

Fig. *Direct Project Cost*

- The straight line or segmented approximation of the direct cost curve is helpful in carrying out the project cost analysis. In such analysis, the cost slope is used.

Cost Slope:

- The cost slope is the slope of the direct cost curve, approximated as straight line. It is defined as follows:

$$\text{Cost slope} = \frac{\text{Crash cost} - \text{Normal cost}}{\text{Normal time} - \text{Crash time}} \quad \text{or} \quad C_S = \frac{C_c - C_n}{t_n - t_c} = \frac{\Delta c}{\Delta t}$$

- The segmented approximation of cost curve, having multiple cost slopes, is more accurate but calculations involved are more. Generally, single cost slope is assumed.

Total Project Cost and Optimum Duration

- The total project cost is the sum of the direct cost and the indirect cost.
- We find that the minimum total cost is obtained at duration known as the optimum duration. The corresponding cost is known as the minimum cost. If the project duration is increased, total cost will increase, while if project duration is decreased to the crash value, project cost will be the highest.

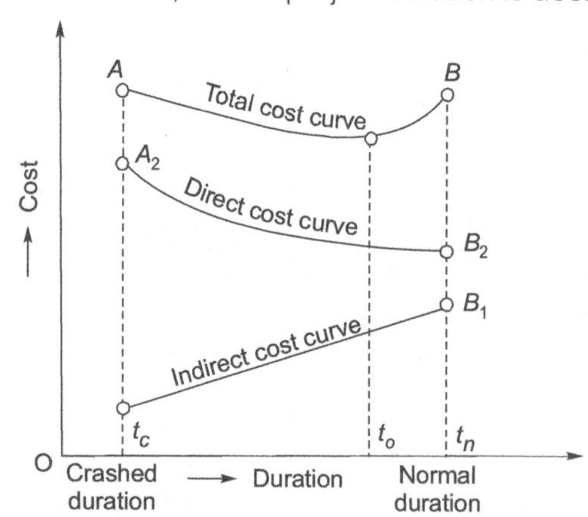

Fig. *Total Project Cost and Optimum Duration*

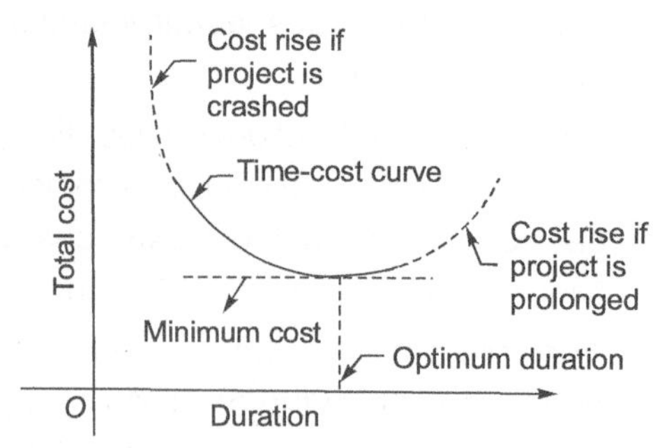

Fig. *Cost-Time Relationship of a project*

Steps in Crashing

1. Estimate project duration and find the critical path.
2. Find cost slope of all activities.
3. The critical activity having minimum cost slope is crashed in first stage. The next stage crashing will involve activity having second lower cost slope in critical path.

4. Total cost of project at this stage is calculated.
5. Steps 3 and 4 are repeated till all activities of project are crashed along critical paths and corresponding time is crash time of project.
6. It is to be noted that only critical activities should be crashed.

Models in Estimating Projects

One of the models used to estimate costs of software projects is called the Constructive Cost Model (or COCOMO) and was developed by Barry Boehm in the late 1070s. Software development effort and cost is estimated by fitting parameters into a regression formula using data from historical projects. This estimation is done based on thousands of lines of source code to be developed on the project.

DO YOU KNOW?

- The certain non-critical activities may also become critical in the process of crashing critical activities.
- Resource smoothening and resource levelling both are trial and error methods.
- Crashing of non-critical activities does not serve any purpose as they do not control the project duration and completing them earlier does no benefit rather it increases the project cost.
- In case of formation of new critical path at any stage of crashing, we have to do parallel crashing of combination of activities from each path for which cost slope is minimum.

EXAMPLE 5.5

Table below shows information about a two activity network. The project overhead costs are Rs. 400 per day. Work out the least cost plan for the network.

Activity	Normal duration (days)	Normal Cost (Rs.)	Crashing duration (days)	Crash Cost (Rs.)
1-2	6	7000	4	9000
2-3	10	9000	5	13000

Solution:

In the network shown, normal duration of activities are marked adjacent to arrow and corresponding crash duration in brackets.
The cost slope of each activity is computed below:

Activity	ΔC (Rs.)	Δt (days)	Cost slope = $\frac{\Delta C}{\Delta t}$ (Rs./day)
1-2	9000 – 7000 = 2000	6 – 4 = 2	1000
2-3	13000 – 9000 = 4000	10 – 5 = 5	800

Total normal duration of the project = 6 + 10 = 16 days.
Total normal cost incurred on the project = 7000 + 9000 = Rs. 16000/-
Now activity 2-3 has the minimum cost slope. Thus crash activity 2-3 first. The duration of activity 2-3 be crashed by 10 – 5 = 5 days.
∴ Extra cost incurred in crashing the activity 2-3 = 800 × 5 = Rs. 4000/-

After crashing activity 2-3, revised duration of project = 6 + 5 = 11 days.

∴ Direct cost of project for 11 days duration

= Normal cost of project + Extra cost incurred in crashing the activity 2-3

= 16000 + 4000 = Rs. 20000/-

After crashing activity 2-3 fully, now crash activity 1-2 from its normal duration of 6 days to crash duration of 4 days i.e., crashing activity 1-2 by 2 days.

Extra cost incurred in crashing the activity 1-2 = 1000 × 2 = Rs. 2000/-

∴ Revised project duration = 4 + 5 = 9 days.

∴ Direct cost of project = 20000 + 2000 = Rs. 22000/-

Duration (days)	16	11	9
Direct cost (Rs.)	16000	20000	22000
Indirect cost (Rs) Rs. 400 per day	16 × 400 = 6400	11 × 400 = 4400	9 × 400 = 3600
Total cost (Rs.)	22400	24400	25600

Thus total cost in minimum for project duration of 16 days and corresponding minimum cost is Rs. 22400/-

EXAMPLE 5.6

For the network shown below, determine the optimum duration and optimum cost.

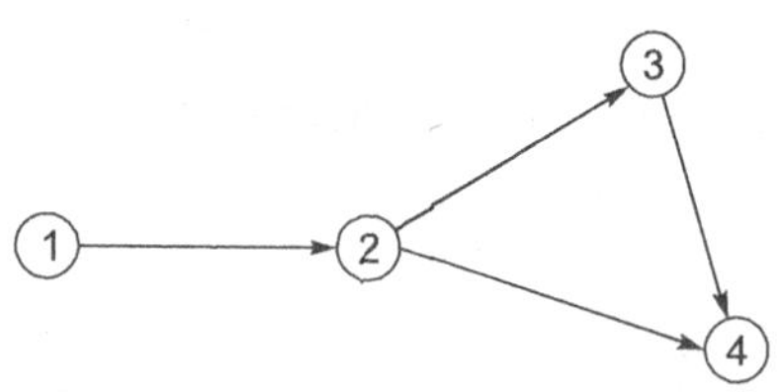

Activity	Normal time (days)	Cost time (days)	Normal cost (Rs.)	Crash Cost (Rs.)
1-2	6	3	500	650
2-3	7	4	250	350
2-4	3	1	350	525
3-4	5	3	700	950

Indirect cost on the project occurs @ Rs. 105 per day.

Solution:

The cost slope of each activity is computed and shown below:

Activity	Normal time (days)	Cost time (days)	Normal cost (Rs.)	Crash Cost (Rs.)	Cost slope (Rs./day)
1-2	6	3	500	650	50
2-3	7	4	250	350	33.33
2-4	3	1	350	525	87.5
3-4	5	3	700	950	125

Network with normal duration activity times

$$\text{Time duration} = 18 \text{ days}$$

$$\text{Total cost} = (500 + 250 + 350 + 700) + (105 \times 18)$$

$$= \text{Rs. } 3690/-$$

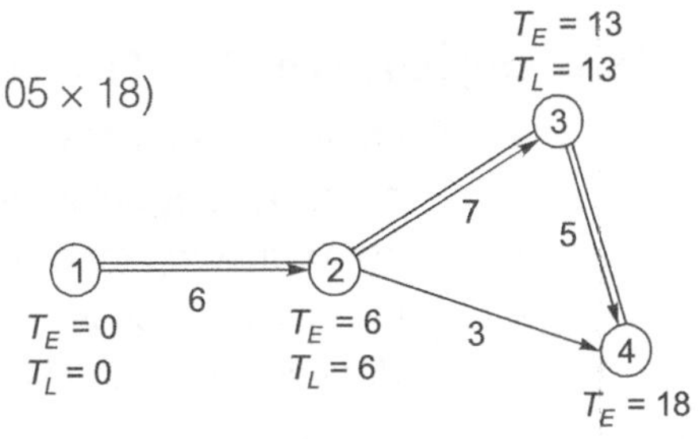

Now, activity 2–3 has minimum cost slope

So crashing activity 2–3 by 3 days.

Hence revised duration of project = 18 – 3 = 15 days

$$\text{Indirect cost} = 105 \times 15 = \text{Rs. } 1575$$

$$\text{Direct cost} = 500 + 250 + 350 + 100 + (33.333 \times 3) = \text{Rs. } 1900$$

Crashed network

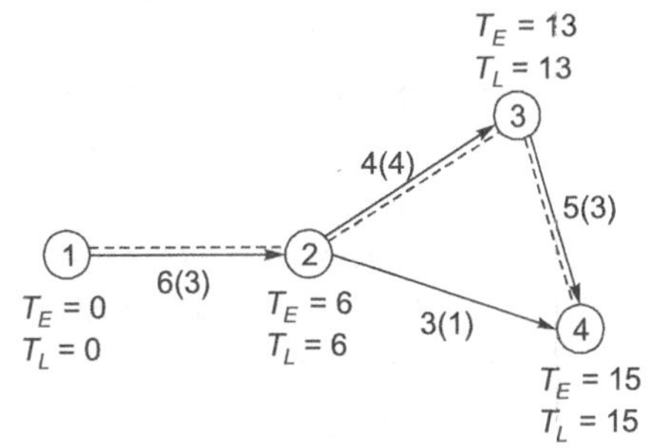

Still, the critical path is 1–2–3–4

Now, activity 1–2 has minimum cost slope on this critical path.

So, crashing 1–2 by available 3 days

Hence,

$$\text{Revised duration} = 12 \text{ days}$$

$$\text{Indirect cost} = 105 \times 12 = \text{Rs. } 1260$$

$$\text{Direct cost} = 1900 + (50 \times 3) = \text{Rs. } 2050$$

$$\text{Total cost} = \text{Rs. } 3310$$

Crashed network

Still critical path is 1–2–3–4

One activity (3–4) along critical path remains to be crushed.

Even if activity 3–4 is crashed by available 2 days.

The critical activities will not change.

So crashing 3–4 activity by 2 days.

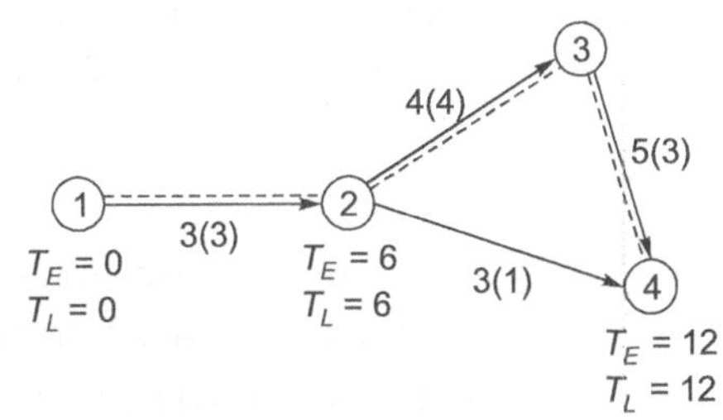

Hence,

$$\text{revised duration} = 12 - 2 = 10 \text{ days}$$

$$\text{Indirect cost} = 105 \times 10 = \text{Rs. } 1050$$

$$\text{Direct cost} = 2050 + (125 \times 2) = \text{Rs. } 2300$$

$$\text{Total project cost} = \text{Rs. } 3350$$

Now, all the activities along critical path are crushed.

Further crashing not required.

Duration (days)	18	15	12	10
Direct cost (Rs.)	1800	1900	2050	2300
Indirect cost (Rs) Rs. 400 per day	1850	1575	1260	1050
Total cost (Rs.)	3690	3475	3310	3350

Thus is can be seen that for optimum duration

$$t = 12 \text{ days}$$

$$\text{Minimum cost} = \text{Rs. } 3310$$

5.15 PROJECT COST ESTIMATE

5.15.1 Objectives of Cost Estimate

- To create an estimate of anticipated costs for individual activities on the project from beginning to end
- To create a budget for the project by aggregating the estimates for individual activities
- To obtain funding for project based on this budget
- To use this budget as a management and control tool
- Under-estimation and Over-estimation – Both are bad for the project

5.15.2 Methods of Cost Estimate

There are four methods to estimate durations and costs related to project activities:

- Analogous Estimate (Top-Down Estimate)
  Estimates are created by using past experience.
- Bottom-Up Estimate
  Individual items are estimated separately by creating a WBS and then rolled up for a project total. Done on projects that are totally new to us.
- Parametric Estimate
  For all parameters of the project like manpower, materials etc.
  Parameters are established for creating the estimate.
  Also created from past experience.
  Also used for R&D projects estimates are worked out based on the rates of some past project of similar kind. Generally a formula is worked out based on the past date.
- Computerized/Monte Carlo Estimate
  Software is used to estimate individual items and then rolled up for a project total.

Top-Down and Bottom-Up Estimates

- High level estimates are often referred to as top-down, and detailed estimates are referred to as bottom-up. The BS technique enables the project manager to estimate at different levels of details. The estimate can be either a higher level (top-down) for the whole project or a more accurate bottom-up estimate at the work package level. At the bottom level the costs can be further subdivided into the different cost types.

  A top-down estimate for a house project can be estimated very quickly and simply as $1,000 per sq. m. using typical square metre rates. A bottom-up estimate for the same project would aggregate all the individual contracts; builder, plumber, electrician, materials, plant hire, etc. The bottom-up estimate should be more accurate, but it will take longer to compile and will incur estimating costs.

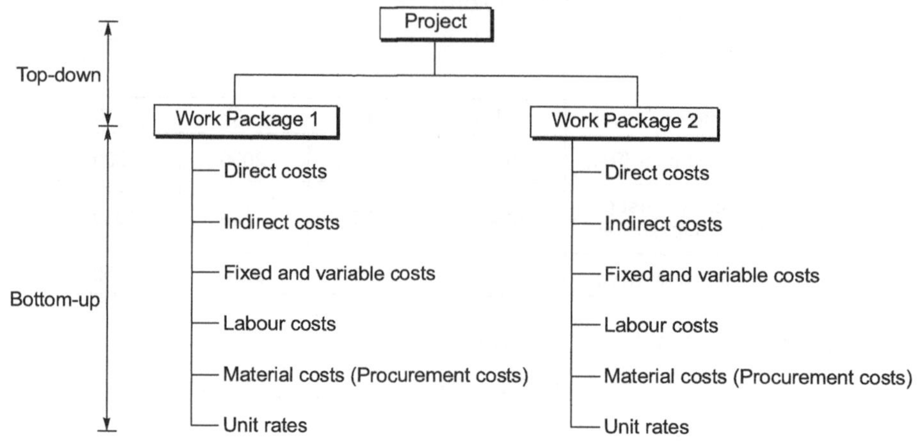

Fig. *Top-Down and Bottom-Up Estimating*

Comparison of the Estimating Method

Method	Advantage	Disadvantage
Analogous (or Top-Down)	Can be created quickly	Lacks detail regarding individual items
Bottom Up	Details and accuracy associated with it.	• Can take significant time to create • Team can pad the estimates to compensate for unknowns.
Parametric	• Can be done quickly • Usually accurate	Lacks detail regarding individual items
Computerized /Monte Carlo	• Accuracy of estimates • "What-if" analysis can be performed.	• Ramp-up time and costs associated with the setup of the tool • Needs an experienced operator for the tool.

5.15.3 Project Cost Components

Total cost of a project includes:

- Land cost
- Plant and machinery
- Transport and erection cost
- Contingency fund
- Margin money for working capital
- Building cost
- Electrical items
- Know-how/consultancy
- Miscellaneous

5.15.4 Cost Funnel

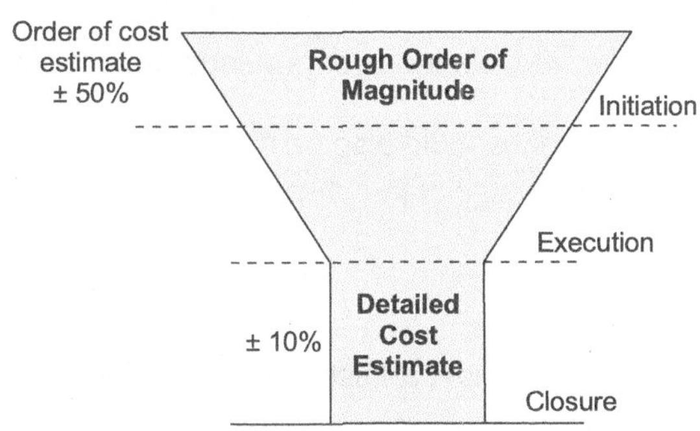

5.16 TYPES OF COST ESTIMATES

5.16.1 Conceptual Cost Estimate

- It is also known as rough estimate, preliminary estimate, feasibility estimate, analogy estimate. It is generally considered as basis for all the cost estimate.
- Usually it is done by owner, it is based on cost per unit item. It is done when construction details available are less than 30% and it may be error of 10 to 20%.

There are three methods for conceptual cost estimate:

1. Unit cost method
2. Factor method
3. Simulation method

1. **Unit Cost Method:** Unit cost is generally prepared using previous projects data, hence it is necessary that estimator should have cost per unit for previously same type of completed project.

$$\text{Unit cost} = \frac{A+4B+C}{6}$$

Also called as weighted unit cost.

A = Minimum unit cost of previous project

B = Average unit cost of previous project

C = Maximum unit cost of previous project

Following adjustments are needed in unit cost:

(a) **Adjustment of time:** The adjustment between two projects represents the relative inflation and deflation of cost, material rate, interest rate.

$$\text{Adjustment cost} = \left(\frac{\text{Cost index for current year}}{\text{Cost index for the year of sample project}}\right) \times \text{Cost of previous project}$$

(b) **Adjustment of location:** Tender prices vary according to the region of the country. Similarly use of cost information from previous project to forecast the cost of a proposed project will not be reliable unless an adjustment is made proportional to represent the difference in cost between locations of two projects.

Construction cost of project @ city A

$$= \left(\frac{\text{Cost index for city A}}{\text{Cost index of previous project city}}\right) \times \text{Cost of previous project}$$

(c) **Adjustment of size:** Adjustment should also be made for difference in the size of two projects. The adjustment is simple ratio of the size of the proposed project to size of similar project for which cost data is available.

(d) **Combined adjustment:**

Proposed cost = Previous cost × Time adjustment × Location adjustment × Size adjustment

EXAMPLE 5.7

Use the weighted unit cost to determine the conceptual cost estimate for a proposed project of parking lot that is expected to contain 135 parked cars. Previous data of similar project is given.

Project No.	Cost (Rs.)	No. of Carb	Unit cost (Rs./Unit)
1.	4,66,580	150	3110.53
2.	2,90,304	80	3628.81
3.	5,25,096	120	4375.8
4.	3,49,920	90	3888.0
5.	2,90,290	60	4321.5
6.	6,57,206	220	2987.3
7.	2,91,718	70	4167.4
8.	7,11,414	180	3952.0

Solution:

$$\text{Forecast weighted unit cost} = \frac{A + 4B + C}{6}$$

$$A = \text{Rs. } 2987.3$$

$$B = \frac{3110.53 + 3628.81 + 4375.8 + 3888 + 4321.5 + 2987.3 + 4167.4 + 3952}{8} = \text{Rs. } 3803.9$$

$$C = \text{Rs. } 4375.8$$

$$UC = \frac{4378.8 + 4 \times 3803.9 + 2987.3}{6}$$

$$UC = \text{Rs. } 3763.1$$

$$\text{Project cost} = UC \times \text{No. of units}$$

$$= 3763.1 \times 135 = \text{Rs. } 508018.5$$

5.16.2 Semidetailed Cost Estimate

Usually it is done by engineer to prepare tender notice:

It is suitable when 30% to 80% construction documents are available.

It may have 5 to 10% error.

5.16.3 Detailed Cost Estimate

It is done by contractor in which quantity of item is multiplied by the cost of that item.

It is suitable when 80 to 100% drawings and specifications are available.

It may have 2 to 4% error.

It is most accurate approach.

Methods of Detailed Cost Estimate:

1. Individual wall method/separate wall method/short and long wall method
2. Centre line method

5.17 PROJECT SUPPORT PLAN

The facilitating processes, are general interactions among the planning processes that are dependent on the nature of each project. Although these processes are performed intermittently and in no particular sequence during project planning and execution, they are not optional and include six additional components of a project plan:

Quality planning: You should identify the quality standards relevant to the project and determine how to achieve them.

Organizational Planning – WBS (Work Breakdown Structure): Here, you should identify, document and assign project roles and responsibilities among the staff and set the reporting relations. The next step is Staff Acquisition, or getting the needed human resources assigned to and working on the project. A useful tip is to form clear and manageable work packages that correspond to the individual work performed by each of the staff and create a document which will serve as a reference point for managing project progress.

Setting a Communications plan: This involves configuring the information and communications needs of all the stakeholders (who needs what information, when will they need it and how they will get it).

Risk Management Planning, is one of the most important components of a project plan and decides the approach and plan for risk management in a project. In other words, it is the safety net of each project and involves the following processes: Identifying key risks likely to affect the project and documenting the characteristics of each. Performing a Qualitative Risk Analysis of the project's risks and conditions in order to prioritize their effects on the project objectives. Running a Quantitative Risk Analysis, measuring the probability and impact of each of the risks and estimating their impact on the project's objectives. Develop a Risk Response Plan, building up a mechanism to strengthen contingencies and reduce any threats to the project's objectives from risk.

Procurement Planning: Defining what to procure, how much to procure and when.

Solicitation Planning: Documenting product requirements and identifying potential sources.

All the project management plan components listed above are to be gathered together to form a cohesive and coherent document that includes all the phases of the life cycle of the project. Hence, it is essential to share this document with all the stakeholders of the project, to read through it, communicate any unclear points, exchange opinions, negotiate and make all the necessary alterations so as to share a finalized version with all involved.

Once your project plan is complete, the document can be used to resolve issues and identify your progress as the project moves forward. In other words, a good plan is essential for delivering a successful project.

5.18 PROJECT RISK

Risk is an event that has some degree of uncertainty attached to it.

A key objective of performing project risk management is to increase the probability of project success by minimizing or eliminating negative risk events and increasing the probability of project success by minimizing or eliminating negative risk events and increasing the probability of positive events.

Risk can be:

Positive Risk – offering benefit to the project

Negative Risk– creating loss to the project

- Risk management includes the processes of planning risk management, identifying risks, performing qualitative and quantitative analysis of risks, planning responses to risks, and finally monitoring and controlling risks on the project.
- Risk is always in the future
- Risk involves uncertainty-it may or may not occur
- Assess your organization's tolerance for risk : are they risk tolerant or risk averse?
- There will be known risks and unknown risks.
- Risks can be negative or positive and are identified as threats or opportunities
- Risk is considered from the moment the project is conceived

Difference between Risk and Uncertainty

"Risk" can be defined as the variability of return from an investment. If the probabilities of possible outcomes of a given problem are known, we can conclude that the problem contains risk: in other words the problem is risky. On the other hand, if the probabilities of possible outcomes of a given problem are, not known, we can conclude that the problem has an element of uncertainty, *i.e.,* the outcome can not be predicted.

5.18.1 Type Project Risks

Some important project risks are as under:

- **Project Completion Risk:** Completing the project in time and within the estimated cost itself is a major achievement. A project that is delayed will result in time over-run which will consequently result in cost over-run.
- **Resource Risk:** Raw material, power, fuel, manpower etc. are the resources used by a project. Shortage of raw material may lead to reduction in capacity utilization and higher cost of production.

- **Price Risk:** Price fluctuations of both inputs and outputs *(i.e..* raw material and finished products) affect the project..
- **Technology Risk:** Technology risk may appear in two forms.
    1. A project that is based on unproven technology *(i.e.,* a technology that is proved at laboratory level but not proved at commercial level) may have hidden defects which may make the project a non-starter.
    2. Rapid growth in technology may make a project obsolete in technology due to the evolution of latest technology.
- **Political Risk:** The Government intervenes in many forms such as levying and regulating taxes, regulating monopolistic trade practices, imposing import duties, promoting exports, prohibiting export of certain commodities, issuing import licenses, controlling foreign exchange transactions, price controls, exploration, nationalization etc. Political risk is a major risk since it can not be predicted easily.
- **Interest Rate Risk:** Fluctuations in interest rate may bring in an adverse effect. Interest rate risks can be managed to some extent by entering into interest rate hedging agreements like 'interest cap', 'interest swap' etc..
- **Exchange Rate Risk:** Exchange rate risk (also called currency risk) is the risk arising from currency fluctuations. Firms exposed to international economy face this risk. When a firm has already committed to a foreign currency denominated transaction. Many exchange rate risk hedging tools such as forward cover, leads and lags, currency options, currency swaps etc., are available which can be efficiently made use of to manoeuvre exchange rate risk.
- Market risk and quality risk

No.	Positive Risk	Negative Risk
1.	Exploit the opportunity	Avoid the situation from happening
2.	Share the opportunity with third experienced party so to capture the opportunity	Transfer of some aspects to third party
3.	Enhancing the changes and outcome of positive risk	Mitigating the chances and outcome of negative risk
4.	Accepting the opportunity without actively pressuring it.	Accept when avoid, transfer and mitigation not possible and there is no well defined strategy. Hence either you become promise or active (keep contingency reserve)

5.18.2 Risk Management or Managing the Risk

1. **Risk reduction strategies:** Change the proportion of fixed and variable costs. [Buy components from outside]
2. **Change pricing strategy:** A lower price (selling price) increases demand but increases break even level.
3. **Sequential investment:** Initially start with low production/small unit and increase the production with increase in demand.
4. **Improving information:**
   - Gather more information about market and technology before taking a plunge.
   - "Don't test the depth of river with both feet"
5. **Financial leverage:** Reduce the dependence on debt and increase equity.
6. Provide insurance cover
7. Enter into long term agreements with suppliers, employees, lenders and customers.
8. Strategic alliance and tie-ups.
9. Explore reinforcing options.

5.18.3 Risk Information Gathering Techniques for Risk Identification

The above-mentioned risk information gathering techniques are well established and have been in use for many years for identification of risk by the project management.

- Brainstorming
- Root Cause Analysis
- SWOT Analysis
- Delphi Technique
- Expert Interviewing

Brainstorming

The four fundamental steps to brainstorming are as follows :

1. **Focus on quantity :** we assume more no of ideas increases solution probability.
2. **Withhold criticism :** In brainstorming, criticism of ideas generated is postponed till all the ideas are generated so everyone feels free to give new idea.
3. **Welcome unusual ideas :** sometimes unusual ideas do fit the solution.
4. **Combine and improve ideas :** at last all ideas are combined to form the proper working idea.

SWOT Analysis:

Internal to organization	Strength	Weakness
External to organization	Opportunity	Threat

- Strengths - Opportunities – identifies how organizational strengths help the organization to capitalize on opportunities
- Weaknesses-Opportunities-identifies organizational weaknesses, from the perspective of making the necessary improvements, to enable the organization to capitalize on opportunities
- Strengths-Threats-identifies how our organizational strengths help us deal with competitors or threatening situations
- Weakness-Threats-identifies how our organizational weaknesses leave us vulnerable to external competition or threats

Risk Register

The risk register is a document that contains information about identified project risks, analysis of risk severity and evaluations of the possible solutions to be applied. Presenting this in a spreadsheet if often the easiest way to manage things, so that key information can be found and applied quickly and easily. Risk register is a living document that needs continuous updates.

The risk register addresses risk management in four key steps:

(i) Identifying the risk,

(ii) Evaluating the severity of an identified risks,

(iii) Applying possible solutions to those risks and

(iv) Monitoring and analysing the effectiveness of any subsequent steps taken.

It contains the following information

1. Major Risks in the projects
2. Triggers that help identify when risk may occur
3. Results from Risk analysis.
4. Responses to those risks events should they occur
5. Risk owners, i.e. who will be responsible & take action in event of their occurrence.
6. Contingency Plans -these are plan B for any identified risk.

Risk Priority	Description	Probability	Impact	Risk Category	Risk Trigger	Response data	Contingency plan	Owner

Any risk whose probability becomes more than 70% is termed as issue.

5.19 PROJECT RISK ANALYSIS

- Risk fearers to variability.
- In project appraisal, risk analysis play an important part. All projects are prone to some kind of risk or the other. All projects are appraised making certain assumptions.

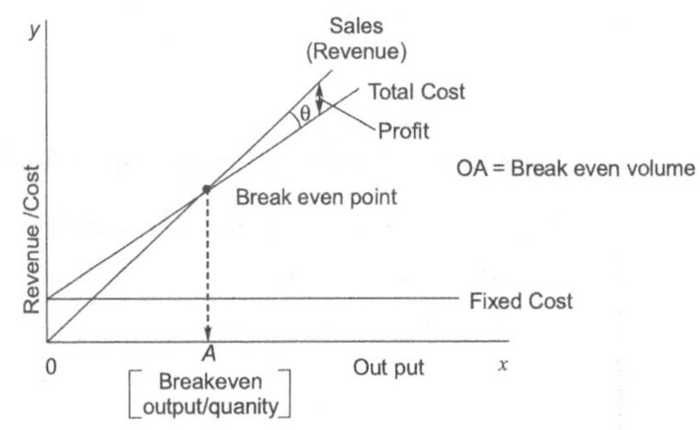

- Though there may be similarity between two different projects, exactly identical projects do not exit. There are many elements of a project which contain assumed values. The following list of elements will give an idea of the assumptions on which any project appraisal is based on.
- Periodic cash inflows
- Periodic cash outflows
- Life of machinery
- Salvage value of machinery

  Since a project is appraised making use of such assumptions, the appraisal is very much prone to the risk of yielding results that may deviate from reality.

Techniques of Risk Analysis

Though there are many mathematical techniques available for risk analysis, the following are the simple tools that come handy for analyzing small and medium sized projects.

1. Sensitivity analysis
2. Scenario Analysis
3. Monte-carlo technique (Simulation Analysis)
4. Break Even analysis
5. Decision tree analysis
6. Game theory

1. Sensitivity Analysis ('What if' Analysis)

- If sales or investment deviates from its expected value then sensitivity analyse is done.
- This analysis indicates that how vulnerable a project is.
- NPV or IRR or any other indicator of project is analysed for variability due to change in any one variable at a time keeping other factor constant.
- The pessimistic, optimistic and expected forecast are made for NPV. Based on it due to change in a variable, the change in NPV is calculated.

Shortcoming of This Method:

- It mearly shows change in NPV/IRR due to changes in some variable.
- In this analysis only one variable is changed at a time where as in real situation several variables tend to more together.
- Interpretation is very subjective one decision maker may accept where as the other may reject the project, based on this analysis.

2. Scenario Analysis

- Most commonly three scenario are considered.
  (i) Optimistic scenario
  (ii) Pessimistic scenario
  (iii) Most likely scenario
- This method is an improvement over sensitivity analysis, because it considers variation in several variables together which is more realistic.

3. Simulation Analysis (Montecarlo Method)

- Sensitivity analysis indicates the variability of NPV, IRR or any other indicator due to variation of any one parameter.

  For ex. If the quantity produced and sold decreases by 1% (other things remain unchanged) then NPV falls by 6% (say). Such information is though useful but not adequate to take decision of investment.
- The decision maker would also like to know the likelihood of such other occurrences. This information can be generated by simulation analysis.

Steps Involved in Simulation Analysis:

(i) **Model the project:** The modeling of project will reveal that how NPV is related to other parameters or exogenous variables.

(ii) Specify value of parameters and probability distribution of exogenous variables.

(iii) Select the value, at random from the probability distribution of each exogenous variables.

(iv) Determine NPV corresponding to randomly generated values of exogenous variables.

(v) Repeat (iii) and (iv), a number of times to get a large number of simulated NPV.

(vi) Plot the frequency distribution of NPV.

Some Commonly Used Probability Distributions are:

Uniform distribution

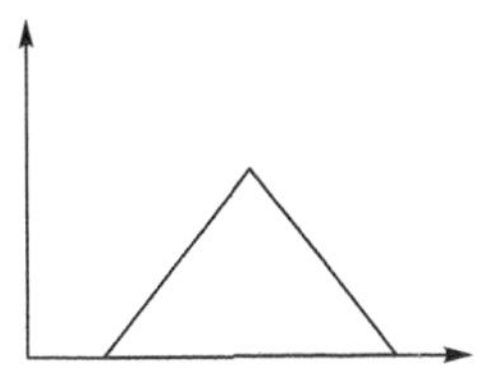

Triangular distribution

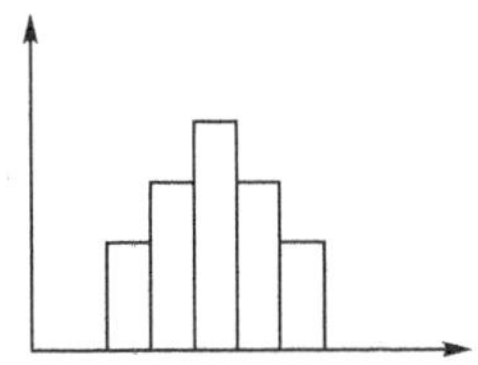

Step-rectangular distribution

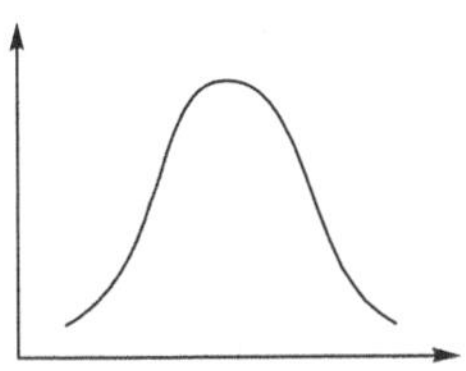

Normal distribution

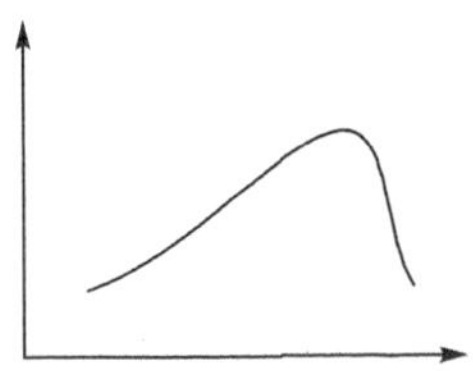

β-distribution

4. Decision Tree Analysis

- It is assumed that the project will be implemented in accordance with a set plan. In reality firms have some flexibility in the execution. When such flexibility exists, decision tree analysis is a useful tool in valuing project flexibility.
- To analyse the situations where sequential decision making has risk involved then decision tree analysis is a useful tool.

Step in Decision Tree Analysis:

1. Identifying the problem and alternatives.
2. Delineating the decision tree (Exhibiting the anatomy of decision situation). It is in a diagrammatic format.
3. Specifying the probabilities and monetary outcomes.
4. Evaluating various decision alternatives.

Risk Response Table

Positive Risk	Negative Risk
Exploit the positive risk to gain from opportunity	Avoid the upcoming risk
Share risk with third party who can fame it better, so as to gain from opportunity	Share risk with third party who can fame it better, so as to gain from opportunity
Work for the enhancement of positive risk	Try to mitigate the effect of negative risk
Accept positive risk to take advantage of opportunity	Accept the negative risk occurrence and respond wisely.

Previous ESE Prelims Questions

Q.1 In a project there are 9 activities : *A, B, C, D* which are sequential; *E, F, G* which are sequential; *H, K* which are sequential. Also *E, F, G* run parallel to *B, C, D*, and *H, K* run parallel to *A, B, C, D*. Besides these activity dependencies. It is also needful that *B* be completed before taking UP G; A and H be completed before taking up *D* and *K*. How many dummies are to be drawn on the activity network?

(a) 5 (b) 4
(c) 3 (d) 2

[ESE-2018]

Ans. (c)

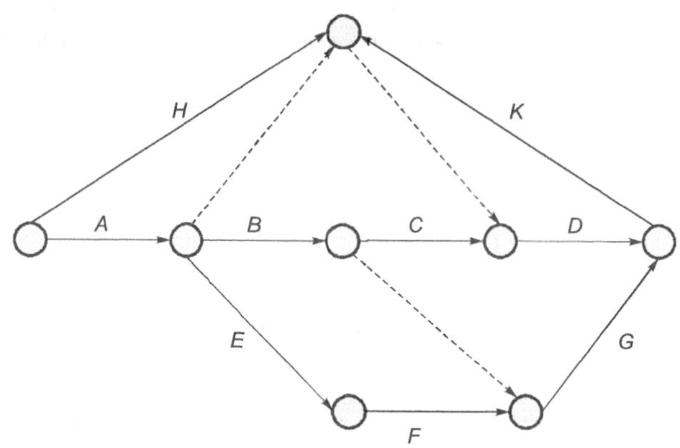

Number of dummies needed = 3

Q.2 Consider the following statements :

1. In work breakdown structure, top-down approach is adopted.
2. Duration along critical path is the shortest duration permissible.
3. PERT is probabilistic in its approach.

Which of the above statements are correct?

(a) 1 and 2 only (b) 1 and 3 only
(c) 2 and 3 only (d) 1, 2 and 3

[ESE-2018]

Ans. (d)

Project will be completed only when activities along critical path are completed. It means minimum project duration is duration of critical path it means shortest duration permissible of project is critical path duration.

Q.3 If L_i = the latest occurrence time for event *j*.
E_i = the earliest occurrence time for event *i*,
T_{ij} = the duration of activity $\vec{ij}$?

(a) $E_i - L_j - T_{ij}$ (b) $L_j - L_i - T_{ij}$
(c) $T_{ij} - E_i - L_j$ (d) $T_{ij} - L_j - E_i$

[ESE-2018]

Ans. (b)

$$F_T = T_{L_J} - T_E^i - t_{ij}$$
$$= L_j - E_i - T_{ij}$$

Q.4 **Statement (I):** Preparation of bar charts is merely a scheduling operation while the preparation and analysis of a network is a planning function.

Statement (II): A bar chart, prima facie, does not show the interrelationships between activities.

(a) Both Statement (I) and Statement (II) are individually true and Statement (II) is the correct explanation of Statement (I).

(b) Both Statement (I) and Statement (II) are individually true but Statement (II) is **not** the correct explanation of Statement (I).

(c) Statement (I) is true but Statement (II) is false.

(d) Statement (I) is false but Statement (II) is true. [ESE-2018]

Ans. (a)

Q.5 CPM method of network analysis is

1. Ideally suited for linearly extending works.
2. Meant essentially for research and development activities.
3. Activity-oriented.
4. Used for planning, scheduling and controlling purposes.

Which of the above statements are correct?

(a) 1 and 2 only
(b) 2 and 3 only
(c) 3 and 4 only
(d) 1 and 4 only [ESE-2018]

Ans. (c)

Q.6 A simple project comprises of two start-to-end parallel paths, each with three activities in series, with no interpath dependencies. The *a*, *m*, *b* data (in days) for each activity are shown in the diagram. Assuming that three activities in series are enough for further computations, what will be the total project duration and its standard deviation?

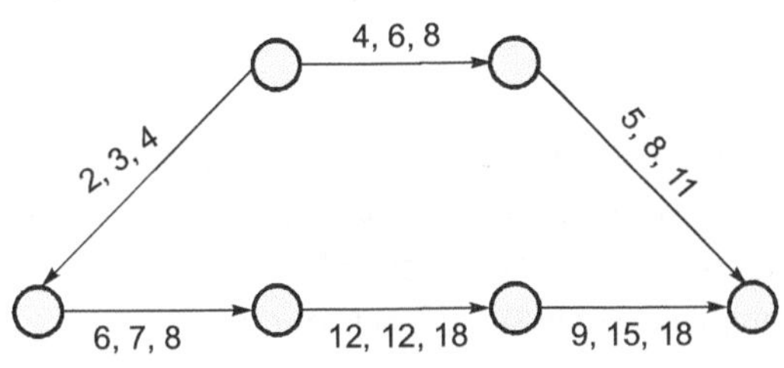

(a) $35\frac{1}{2}$ day and $\frac{14}{3}$ days
(b) $34\frac{1}{2}$ day and $\frac{5}{2}$ days
(c) $35\frac{1}{2}$ day and $\frac{13}{2}$ days
(d) $34\frac{1}{2}$ day and $\frac{11}{2}$ days [ESE-2018]

Ans. (d)

Project duration along top path

$$= \frac{2+4\times3+4}{6}+\frac{4+4\times6+8}{6}+\frac{5+4\times8+11}{6}$$

$$= 3 + 6 + 8 = 17 \text{ days}$$

Project duration along bottom path

$$= \frac{6+4\times7+8}{6}+\frac{12+4\times12+18}{6}+\frac{9+4\times15+18}{6}$$

$$= 7 + 13 + 14.5 = 34.5 = 34\frac{1}{2}$$

Longest is $34\frac{1}{2}$, hence this path is critical.

$$\sigma = \sqrt{\left(\frac{8-6}{6}\right)^2 + \left(\frac{18-12}{6}\right)^2 + \left(\frac{18-9}{6}\right)^2}$$

$$= \sqrt{\left(\frac{1}{3}\right)^2 + 1 + \left(\frac{3^2}{2}\right)} = \sqrt{\frac{1}{9} + 1 + \frac{9}{4}}$$

$$= \sqrt{\frac{4+36+9\times 9}{36}} = \sqrt{\frac{121}{36}} = \frac{11}{6}\text{days}$$

Q.7 Which one of the following schedules shows the specific activities necessary to complete an activity or work package?

(a) Project schedule
(b) Master schedule
(c) Task schedule
(d) Internal schedule **[ESE-2019]**

Ans. **(c)**

Q.8 In principle, the network should not be made complex. No control system, for that matter, can operate unless it is kept simple. This principle is called

(a) CPM
(b) PERT
(c) KISS
(d) GERT **[ESE-2019]**

Ans. **(c)**

Q.9 Which one of the following risks can be reduced by investing in projects or acquiring other firms that have a negative correlation with the earnings of the firm?

(a) Investment risk
(b) Business risk
(c) Financial risk
(d) Portfolio risk **[ESE-2020]**

Ans. **(d)**

Q.10 Which one of the following is NOT the purpose of the organization breakdown structure?

(a) To provide a framework to summarize organization unit work performance
(b) Do not tie the organizational unit to cost control accounts
(c) Identify organization units responsible for work packages
(d) How the firm has organized to discharge work responsibility **[ESE-2021]**

Ans. **(b)**

- A project Organisation Breakdown Structure (OBS) is a depiction of the project organization arranged to indicate the reporting relationships within the project context. The OBS reflects the way the project is functionally organized. It is a direct representation and description of the hierarchy and organizations that will provide resources to plan and perform work identified in the Work Breakdown Structure (WBS).
- The OBS helps management focus on establishing the most efficient organization, by taking into consideration availability and capability of management and technical staff including subcontractors, to achieve project objectives.
- The OBS is depicted on the Responsibility Assignment Matrix (RAM), where it is used to identify the organization responsible and accountable for every element of the WBS and Scope of Work (SOW).

Q.11 Which one of the following is NOT a condition for preferring Top-Down Time and Cost Estimates?

(a) Strategic decision making
(b) Cost and time important
(c) High uncertainty
(d) Internal, small project **[ESE : 2022]**

Ans. (b)

Top-down estimates is a technique used by upper-level management to estimate the total cost of a project by using information from a previous, similar project.

This approach is practical for the initial stage of strategic decision-making and in situations where the information required to develop accurate duration and costs estimates is not available in the initial phase of the project.

Q.12 In Network Computation Process, which one of the following is correct for forward pass?

(a) It starts with the first project activity(ies) and traces each path (chain of sequential activities) through the network to the last project activity(ies).

(b) This is the longest path in the network, which will delay the project.

(c) It starts with the last project activity(ies) on the network.

(d) It starts with the last project activity(ies) and traces each path (chain of sequential activities) through the network to the first project activity(ies)

[ESE : 2022]

Ans. (a)

Q.13 Which one of the following projects are those which are to be completed within a stipulated time, even at the cost of ending up with a higher project cost?

(a) Normal projects (b) Business projects

(c) Crash projects (d) Research projects [ESE : 2022]

Ans. (c)

Q.14 Consider the following statements for hammock activities:

1. It derives its name because it spans over a segment of a project.
2. The hammock activity duration is determined after the network plan is not drawn.
3. The hammock activities are frequently used to identify the use of fixed resources or costs over a segment of the project.
4. The maximum amount of time an independent activity must be delayed to begin or end.

Which of the above statements are correct?

(a) 2 and 3 only (b) 1 and 3 only

(c) 1 and 4 only (d) 2 and 4 only [ESE : 2023]

Ans. (b)

- A hammock or summary activity is used to gather together a number of sub-activities into one master activity.
- Hammock activities show the included subactivities.
- Hammock activities are frequently used to identify the use of fixed resources or costs over a segment of project.

Q.15 Consider the following strategies for mitigating risk under risk response development:

1. Reduce the likelihood that the event will occur.
2. Reduce the impact that the adverse event would have on the project.
3. Analyze the project to identify sources of risk.
4. Assess risks in terms of severity of impact.

Which of the above strategies are correct?

(a) 1 and 2 only (b) 3 and 4 only

(c) 1 and 4 only (d) 2 and 3 only

[ESE : 2023]

Ans. (a)

As per Krause, 2014, Mitigating risk involves two basic strategies : reducing the likelihood that the event will occur and/or reducing the impact that the adverse event would have on the project.

- The risk response plan development process considers risk transfer, reduction, avoidance and other mitigation tools to develop a risk response plan to ensure that the appropriate risk warning tools are in place, to handle risk efficiency.
- Proper risk management will attempt to reduce the probability of an event occuring and/or the magnitude of its impact as well as increase the probability of project success.

Q.16 Consider the following statements:

The strategy is to assign extra time at critical moments in the project, buffers are added to:

1. activities with no risk.
2. merge activities that are prone to delays due to one or more preceding activities being late.
3. non-critical activities to reduce the likelihood that they will create another critical path.
4. activities that require scarce resources to ensure that the resources are available when needed.

Which of the above statements are correct?

(a) 1, 2 and 3 only (b) 1, 2 and 4 only

(c) 2, 3 and 4 only (d) 1, 3 and 4 only [ESE : 2023]

Ans. (c)

A buffer is extra time added to plan/schedule so that a task can be done and delays can be considered. The critical chain method has three different types of buffers, as:

1. **Resource Buffer :** It is inserted just before a critical chain activity where a critical resource is required. It is used to remind the project team that a resource is needed and to finish up prior activities.
2. **Feeding Buffer :** It is inserted as a safety margin in the non-critical chain of a network schedule. It is placed where the path feeds back into the critical chain path.
3. **Project Buffer :** It is the summation of all the internal buffers added to each project task or activity.

Q.17 Which one of the following does NOT always yield an optimal schedule, however it is capable of yielding a "good" schedule for very complex networks having many types of resources?

(a) Algorithm (b) Optimum

(c) Backhoes (d) Heuristics [ESE : 2023]

Ans. (d)

Heuristics do not always yield an optimal schedule, but they are very capable of yielding a "good" schedule for very complex networks with many types of resources. Since, each project is unique, it is good to test several set of heuristics on a network to determine the priority allocation rules that minimize project delay.

Q.18 According to CCPM, using 50/50 estimates will discourage Parkinson's law, the student syndrome, and self-protection from coming into play because there is less "free time" available. What does the abbreviation CCPM stand for?

(a) Control – Chain Project Management (b) Creating – Chain Project Management

(c) Computer – Control Project Management (d) Critical – Chain Project Management

[ESE : 2023]

Ans. (d)

CCPM stands for critical-chain project management. The CCPM aims to eliminate project schedule delays due to uncertainties, overestimating task duration, etc.

Q.19 According to project cost-duration graph, any reduction in project duration means a reduction in
(a) direct costs (b) indirect costs
(c) total costs (d) optimum costs [ESE : 2023]

Ans. (b)

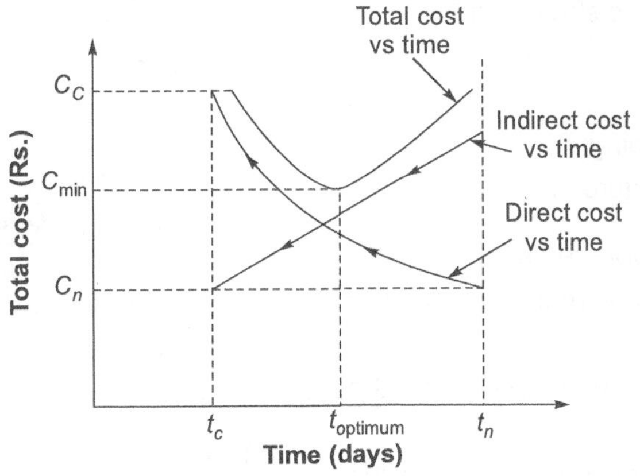

Total cost = Indirect cost + Direct cost

Any reduction in project duration, reduces the indirect cost but increases the direct cost.

Q.20 Which one of the following methods is an important qualitative method under demand forecasting?
(a) Jury of Executive Method (b) Trend Projection Method
(c) Chain Ratio Method (d) Bass Diffusion Method [ESE : 2024]

Ans. (a)

Q.21 Why Delphi method of demand forecasting appeals to many organizations?
(a) The biases underlying are subjective, and it seems to be more accurate and less expensive than the traditional face-to-face group meetings
(b) It is intelligible to users, it is a fancy name, and it seems to be more accurate and less expensive than the traditional face-to-face group meetings
(c) It has immense appeal, the biases underlying are subjective, and it seems to be more accurate and less expensive than the traditional face-to-face group meetings.
(d) It is an expedition's method, it has immense appeal, and it seems to be more accurate and less expensive than the traditional face-to-face group meetings.
[ESE : 2024]

Ans. (a)

Objective Brain Teasers

Q.1 Your company, which operates one of the regions largest chemical processing plants, has been convicted of illegally dumping toxic substances into the local river. The court has mandated that the required cleanup activities be completed by February 15. This date is an example of:
(a) A key event
(b) A milestone
(c) A discretionary dependency
(d) An external dependency

Q.2 Scheduling helps in
(a) Planning for the project
(b) Financial control of the project
(c) Preparing the estimate for the project
(d) Carrying out the project in an orderly and effective manner

Q.3 To control the schedule, a project manager is reanalyzing the project to predict project duration. She does this by analyzing the sequence of activities with the least amount of scheduling flexibility. What technique is she using?
(a) Critical path method
(b) Flowchart
(c) Precedence diagramming
(d) Work breakdown structure

Q.4 The WBS for the project represents:
(a) All the tangible items that must be delivered to the client.
(b) All the work that must be completed for the project.
(c) The work that must be performed by the project team.
(d) All the activities of the project.

Q.5 Consider the following statements in work breakdown structure.
1. It is a graphical representation of entire programme.
2. The top-down approach to planning is adopted.
3. The down-top approach to planning is adopted.
4. It is suitable for complex projects.

Which of these statements is/are correct?
(a) 1 only (b) 2 and 4
(c) 3 and 4 (d) 4 only

Q.6 Consider the following statements:
1. In work-breakdown structure top-down approach is adopted.
2. Bar-chart depicts interdependencies of activities.
3. Controlling can be better achieved in milestone chart.

Which of these statements is/are correct?
(a) 1 and 3 only (b) 1 and 2 only
(c) 2 and 3 only (d) 1, 2 and 3

Q.7 A small project, consists of seven activities in the activity-on-node diagram as shown in the figure below, in days and the predecessor relationships are shown.

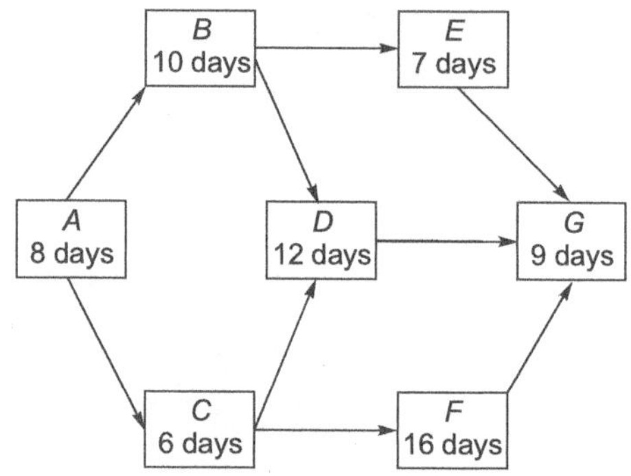

What is the total project duration of the project?
(a) 39 days (b) 35 days
(c) 34 days (d) 41 days

Q.8 What is the proportional variation and what is the 'range' of project duration for the network as shown with the indicated probabilistic *a*, *m*, *b* durations of the respective activities?

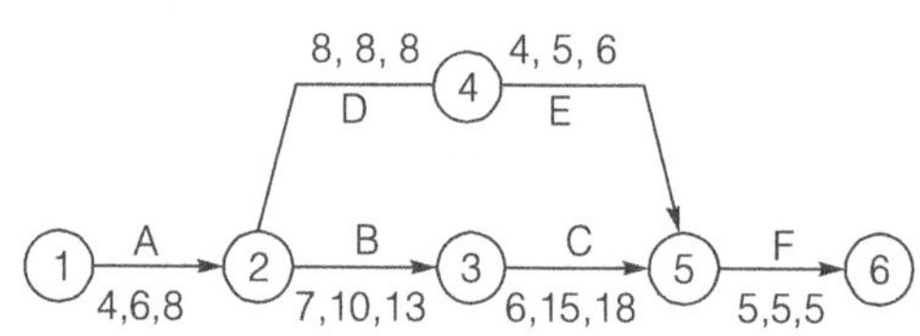

(a) $\frac{1}{15}$, 28 to 42 (b) $\frac{1}{15}$, 30 to 45

(c) $\frac{1}{12}$, 28 to 42 (d) $\frac{1}{12}$, 30 to 45

Q.9 **Statement (I) :** A dummy job takes zero time to perform.

Statement (II) : It is used solely to illustrate precedence relationship.
(a) Both Statement (I) and Statement (II) are individually true and Statement (II) is the correct explanation of Statement (I).
(b) Both Statement (I) and Statement (II) are individually true but Statement (II) is not the correct explanation of Statement (I).
(c) Statement (I) is true but Statement (II) is false.
(d) Statement (I) is false but Statement (II) is true

Q.10 Which of the following are the guidelines for the construction of a network diagram?
1. Each activity is represented by one and only one arrow in the network

2. Two activities can be identified by the same beginning and end events
3. Dangling must be avoided in a network diagram
4. Dummy activity consumes no time or resource

Select the correct answer the codes given below:

(a) 1, 2 and 3 (b) 1, 3 and 4
(c) 1, 2 and 4 (d) 2, 3 and 4

Q.11 Assertion (A): Generally PERT is preferred over CPM for the purpose of project evaluation.

Reason (R): PERT is based on the approach of multiple time estimates for each activity.

(a) both A and R are true and R is the correct explanation of A
(b) both A and R are true but R is not a correct explanation of A
(c) A is true but R is false
(d) A is false but R is true

Q.12 Consider the following statements with respect to PERT:

1. It consist of activities with uncertain time phases.
2. This is evolved from Gantt chart.
3. Total slack along the critical path is not zero.
4. There can be more than one critical path in PERT network.
5. It is similar to electrical network.

(a) 1 and 2 (b) 2 and 4
(c) 1 and 4 (d) 2 and 4

Q.13 Which one of the following statements is not correct?

(a) PERT is probabilistic whereas CPM is deterministic.
(b) In PERT slack on various events in calculation whereas in CPM floats are calculated.
(c) Critical path in a network is the path on which events have no slack.
(d) More than four dummy activities cannot be used in a PERT network.

Q.14 Unlike bar charts, milestone charts show:

(a) Scheduled start or completion of major deliverables and key external interfaces
(b) Activity start and end dates of critical tasks
(c) Expected durations of the critical path
(d) Dependencies between complementary projects

Q.15 An activity has an early start date of the 10th and a late start date of the 19th. The activity has a duration of four days. There are no non-workdays. From the information given, what can be concluded about the activity?

(a) Total float for the activity is nine days.
(b) The early finish date of the activity is the end of the day on the 14th.
(c) The late finish date is 25th.
(d) The activity can be completed in two days if the resources devoted to it are double

Q.16 Several types of float are found in project networks. Float that is used by a particular activity and does NOT affect the float in later activities is called:

(a) Extra float (b) Free float
(c) Total float (d) Expected float

Q.17 Activity A has a pessimistic (P) estimate of 36 days, a most likely (ML) estimate of 21 days, and an optimistic (O) estimate of 6 days. What is the probability that activity A will be completed in 16 to 26 days?

(a) 55.70 percent (b) 68.26 percent
(c) 95.46 percent (d) 99.73 percent

Q.18 Which of the following is the correct sequence to analyze a project for implementation?

(a) Time-cost study, Network, WBS, Scheduling with resource allocation
(b) Network, Time-cost study, Scheduling with resource allocation, WBS
(c) WBS, Network, Scheduling with resource allocation, Time-cost study
(d) WBS, Time-cost study, Network, Scheduling with resource allocation

Q.19 Consider the following statements:

1. A dummy activity is artificially introduced in a network when necessary.
2. A dummy activity consumes some time.
3. A dummy activity is represented by a dotted arrow.
4. A dummy activity must necessarily be introduced in every network.

Which of the above statements are correct?

(a) 1, 2 and 3 (b) 1 and 3
(c) 2, 3 and 4 (d) 1 and 2

Q.20 Match **List-I** (Activity type) with **List-II** (Represented by) and select the correct answer:

List-I

A. Artificially introduced
B. Critical
C. Noncritical type
D. Dangler

List-II

1. A single thick arrow
2. A single arrow
3. An arrow emerging from end event and ending into an event which is not finish event and yet no emerging arrow from that event.
4. A dotted arrow

Codes:

	A	B	C	D
(a)	4	1	2	3
(b)	2	3	4	1
(c)	4	3	2	1
(d)	2	1	4	3

Q.21 Which of the following are critical activities of the bar chart shown below?

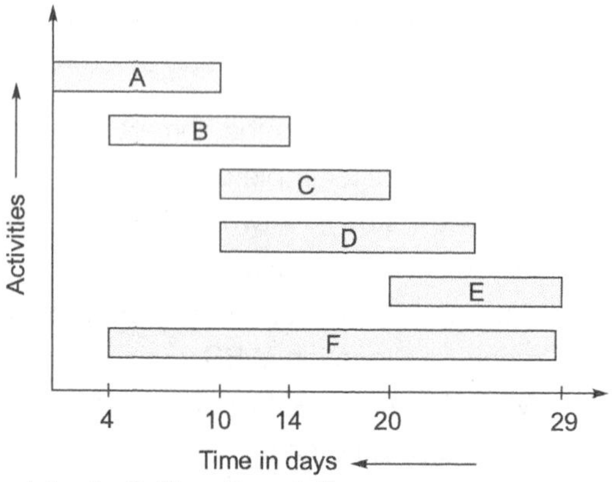

(a) Activities *B* and *E*
(b) Activities *A*, *D* and *F*
(c) Activities *A*, *C* and *E*
(d) Activities *A* and *F*

Q.22 Activity '*C*' follows activity '*A*' and activity '*D*' follows activities '*A*' and '*B*'. The correct network for the project is

(a)
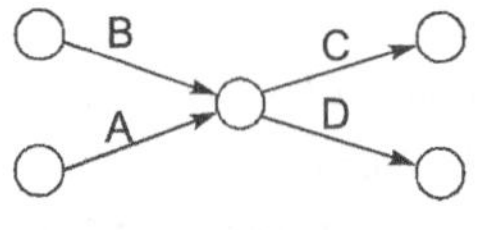

(b)
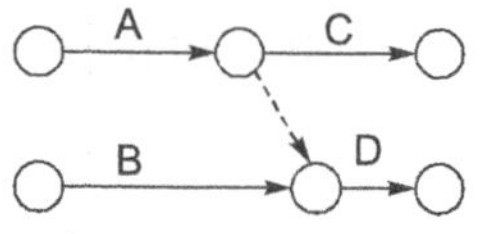

(c)
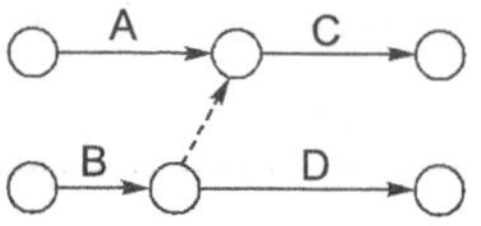

(d)
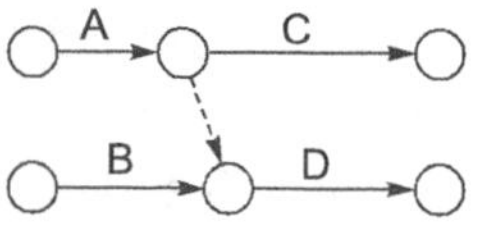

Q.23 The optimistic, likely and pessimistic time estimates for the PERT network of a project are shown in the given figure. The expected duration of the project is

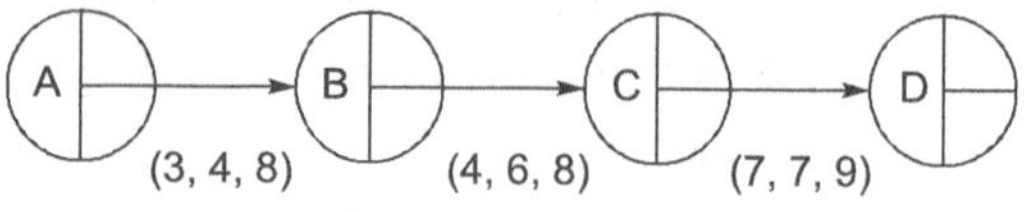

(a) 14 days (b) 17 days
(c) 17.83 days (d) 18.67 days

Q.24 The flow net of activities of a project is shown in the below figure. The duration of the activities are written along the arrows:

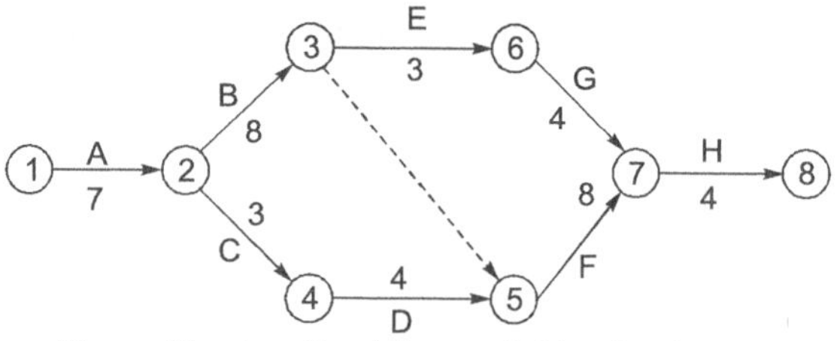

The critical path of the activities is along

(a) 1-2-4-5-7-8 (b) 1-2-3-5-7-8
(c) 1-2-3-6-7-8 (d) 1-2-4-5-3-6-7-8

Q.25 The area under the β-distribution curve is divided into two equal halves by vertical ordinate through

(a) expected time (b) optimistic time
(c) most likely time (d) pessimistic time

Q.26 Consider the following statements of network:

1. Only one time estimate is required for each activity.
2. Three time estimates for each activity.
3. Time and cost are both controlling factors.
4. It is built-up of event-oriented diagram.

Which of the above statements are correctly applicable to CPM network?

(a) 1 and 3 (b) 1 and 2
(c) 2 and 4 (d) 3 and 4

Q.27 In the PERT analysis, which one of the following is followed by the time estimates of activities and probability of their occurrence?

(a) Normal distribution
(b) Poisson's distribution
(c) β-distribution
(d) Binomial distribution

Q.28 What does higher standard deviation imply in cost analysis?
(a) Higher uncertainty
(b) Lower uncertainty
(c) Nothing to do with uncertainty
(d) Extra costs are likely

Q.29 What does the critical path in PERT represent?
1. The shortest path for the earliest completion of the project.
2. The longest path of the network from the initial to final event.

Select the correct answer using the code given below:
(a) 1 only (b) 2 only
(c) Both 1 and 2 (d) Neither 1 nor 2

Q.30 In a PERT network, the activity durations are given as t_0 (optimistic time), t_p (pessimistic time) and t_m (most likely time). What is the variance of the activity?
(a) $\frac{t_0 + 4t_m + t_p}{6}$ (b) $\left(\frac{t_p - t_o}{6}\right)$
(c) $\left(\frac{t_p - t_o}{6}\right)^2$ (d) None of these

Q.31 You want to structure your project so that each project team member has a discrete work package to perform. The work package is a:
(a) Deliverable at the lowest level of the WBS
(b) Task with a unique identifier
(c) Required level of reporting
(d) Task that can be assigned to more than one organisational unit

Q.32 In planning for a project, we should consider the following aspects
(a) The past, present and future
(b) Design considerations
(c) Financial considerations
(d) Man power availability

Q.33 The main objective of planning is to execute the project
(a) Most economically and on time
(b) Only timely completion
(c) Only most economically
(d) Economically and safely

Q.34 Planning is important because it is advantageous to the following
(a) Contractors
(b) Architects and engineers
(c) Clients
(d) All of them

Q.35 One of the main disadvantages of the bar chart is
(a) Progress of the work cannot be monitored
(b) Does not show the interdependencies of the activities
(c) The time schedule is not shown properly
(d) The financial aspect is not shown

Q.36 Work Break down structure helps in
(a) Identifying the activities
(b) Identifying the functional elements of a project and their interrelationship
(c) Breaking the project into several elements
(d) Breaking the project into subsystems

Q.37 Precedence networks are best suited when
(a) There are large number of overlapping activities
(b) There are many critical activities
(c) The project is to be executed efficiently
(d) A-O-A networks are to be avoided

Q.38 The independent float affects
(a) Only succeeding activities
(b) Only preceding activities
(c) Only the particular activity involved
(d) Both the succeeding and preceding activities

Q.39 Interfering float for an activity is the difference between its
(a) Total float and independent float
(b) Free float and independent float
(c) Total float and free float
(d) None of them

Q.40 In the network shown in figure, total float for activity 3-5 is (the number on the arrow shows duration of the activity)

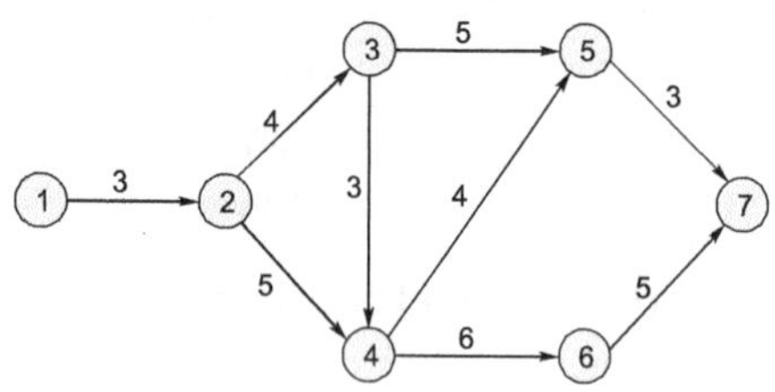

(a) 6 (b) 2
(c) 4 (d) 0

Q.41 In the network shown in figure, the critical path is

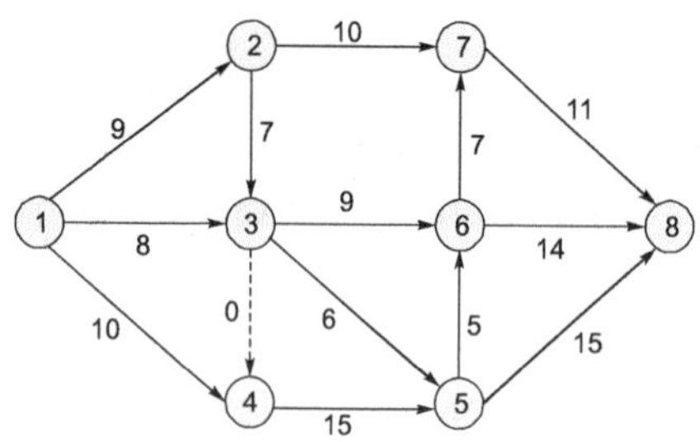

(a) 1-2-3-6-8 (b) 1-2-6-7-8
(c) 1-3-5-6-7-8 (d) 1-2-3-4-5-6-7-8

Q.42 The software package used for the development of networks is
(a) Primavera project planning software
(b) Success planned software
(c) Harvard Total Project Manager (HTPM)
(d) All of these

Q.43 A serious limitation of interdependencies between various activities is generally observed in
(a) bar charts (b) milestone charts
(c) network analysis (d) job layouts

Q.44 Which of the following is the correct sequence to analyze a project for implementation?
(a) Time-cost study, Network, WBS, Scheduling with resource allocation
(b) Network, Time-cost study, Scheduling with resource allocation, WBS
(c) WBS, Network, Scheduling with resource allocation, Time-cost study
(d) WBS, Time-cost study, Network, Scheduling with resource allocation

Q.45 Consider the following tasks:
1. Placing of reinforcement (P) for roof slab cannot start before bending of reinforcement (B) and erection of frame work (E).
2. As soon as placing of reinforcement is finished, concreting (C) will follow.

The correct activity on arrow diagram representing for the above tasks is

(a)
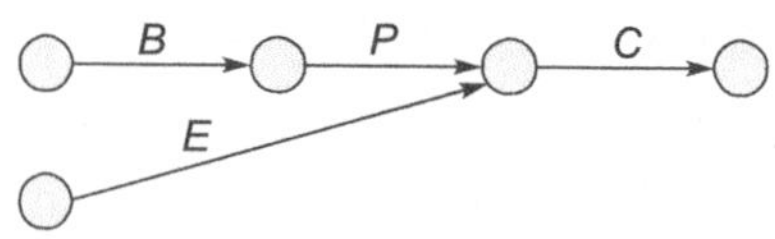

(b)
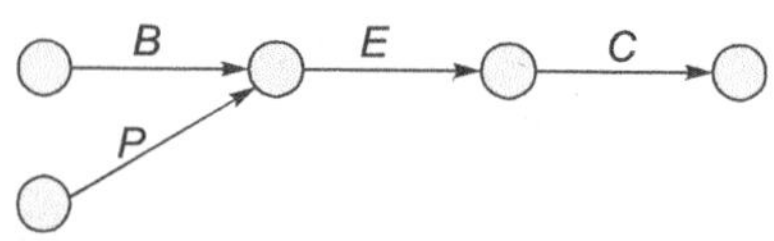

(c)
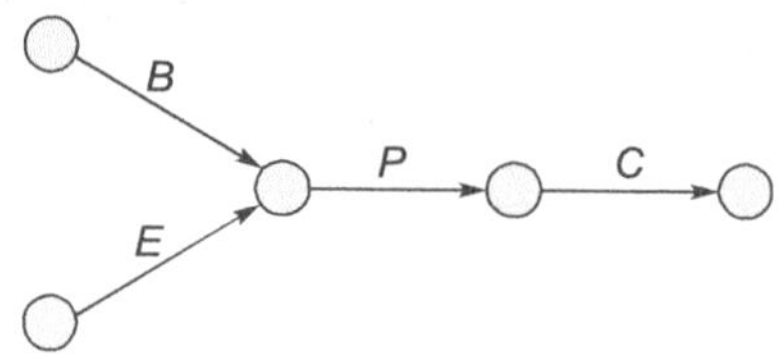

(d)
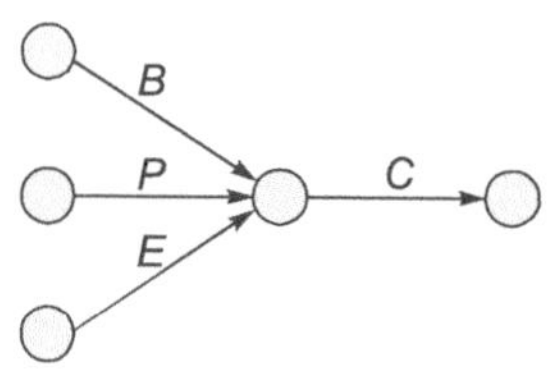

Q.46 Consider the following activities of a housing project:
1. Flooring
2. Wall-plastering
3. Conceal wiring
4. Fixing door-window frames
5. Fixing door-window shutters

What is the correct logical sequence of the above activities?
(a) 4-3-2-5-1 (b) 3-1-5-4-2
(c) 1-4-5-2-3 (d) 1-2-3-4-5

Q.47 Consider the following statements in work breakdown structure.
1. It is a graphical representation of entire programme.
2. The top-down approach to planning is adopted.
3. The down-top approach to planning is adopted.
4. It is suitable for complex projects.

Which of these statements is/are correct?

(a) 1 only (b) 2 and 4
(c) 3 and 4 (d) 4 only

Q.48 **Assertion (A) :** Critical path can pass through dummy arrows also.
Reason (R) : Necessary additional nodes are introduced to ensure uniqueness of *i-j* notation for activities.
(a) Both *A* and *R* are true and *R* is the correct explanation of *A*
(b) Both *A* and *R* are true but *R* is not a correct explanation of *A*
(c) *A* is true but *R* is false
(d) *A* is false but *R* is true

Q.49 What is the time by which the completion of an activity can be delayed without affecting the start of succeeding activities, called?
(a) Total float
(b) Interfering float
(c) Independent float
(d) Free float

Q.50 Slack time is associated with
(a) A real activity
(b) An event
(c) Both event and real activity
(d) Dummy activity

Q.51 In PERT analysis, the time estimates of activities and probability of their occurrence follow:
(a) beta distribution
(b) gamma distribution
(c) normal distribution
(d) Poisson's distribution

Q.52 The work defined at the lowest level of the Work Breakdown Structure to estimate and manage time, risk, quality and cost is called the:
(a) Requirement
(b) Contract
(c) Work package or activity
(d) Cost account

Q.53 Lag means:
(a) The amount of time an activity can be delayed without delaying the project finish date.
(b) The amount of time an activity can be delayed without delaying the early start date of its successor.
(c) Waiting time for an activity.
(d) The product of a forward and backward pass.

Q.54 One of the main reasons to create a schedule for a project is:
(a) For the team to know when to plan their vacations during the project life cycle
(b) To know the end date for the project and to fix the schedule baseline
(c) To know the critical path items to assign resources
(d) To hire the vendors for the project in a planned manner

Q.55 Project milestones have a duration of:
(a) Depends on the type and nature of project
(b) It is decided by the sponsor
(c) The manager and team decide this duration
(d) Zero because it is only a check point on the project

Q.56 A project manager is in the middle of the executing process of a very large construction project when he discovers that the time needed to complete the project is longer than the time available. What is the BEST thing to do?
(a) Cut product scope
(b) Meet with management and tell them that the required date cannot be met
(c) Work overtime.
(d) Determine options for schedule compression and present management with your recommended option.

Q.57 Which of the following generally cannot be validated using a work breakdown structure?
(a) Schedule control (b) Cost control
(c) Quality control (d) Risk management

Q.58 The shortest time necessary to complete all of the activities in a network is called the
(a) Activity duration length
(b) Critical path
(c) Maximum slack path
(d) Compression path

Q.59 A network-based technique that allows for branching and looping is
(a) PERT (b) GERT
(c) PDM (d) ADM

Q.60 You have an activity where the early start is week 6, the early finish is week 10, the latest start is week 14, and the latest finish is week 18. The slack in this activity is

(a) 4 weeks (b) 6 weeks
(c) 8 weeks (d) 18 weeks

Q.61 Most statistical simulations of budgets, schedules, and resource allocations use which one of the following approaches?

(a) PERT
(b) Decision tree analysis
(c) Present value analysis
(d) Monte Carlo analysis

Q.62 In respect to project management plan, which are true

1. It is a well written document always
2. Inputs are given by stakeholders
3. It is created by project manager

(a) All of the above (b) 1 and 2
(c) 1 and 3 (d) 2 and 3

Q.63 **Assertion (A):** Project management plan integrates all the knowledge area i.e. scope, time, cost, quality, procurement, HR, Risk, communication & integration management.

Reason (R): All the knowledge areas are applied uniformly of all projects.

(a) Both Assertion (A) and Reason (R) are individually true and Reason (R) is the correct explanation of Assertion (A).
(b) Both Assertion (A) and Reason (R) are individually true but Reason (R) is NOT the correct explanation of Assertion (A).
(c) Assertion (A) is true but Reason (R) is false.
(d) Assertion (A) is false but Reason (R) is true.

Q.64 Project management plan is needed because

1. To integrate all knowledge areas
2. Performance measurement
3. For execution process of different works in project
4. For monitoring & controlling the changes.

(a) All of the above (b) 1, 2 and 3
(c) 2, 3 and 4 (d) 1, 2 and 4

Q.65 Project management plan includes

1. Baseline for cost, scope, schedule.
2. Scope statement
3. WBS
4. Project manager & Team members

(a) All of the above (b) 1, 2 and 3
(c) 2, 3 and 4 (d) 2 and 3 only

Q.66 A point from where all the subsequent changes to the project are measured and compared is called

(a) kickoff meeting
(b) project charter
(c) performance measurement baseline
(d) Lessons learned

Q.67 Which condition are necessary for kickoff meeting to take place?

1. Phase initiation
2. Project initiation
3. Presence of all stakeholders

(a) All of the above (b) 1 and 2
(c) 1 and 3 (d) 2 and 3

Q.68 Which of the following Techniques are used for scheduling?

1. Ban chart 2. Milo-stone chart
3. PERT and CPM 4. Line of Balance

(a) All of the above (b) 1, 2 and 3
(c) 2, 3 and 4 (d) 1 and 2 only

Q.69 Which of the following is true for a critical path?

1. It is max time consuming path
2. It is shortest project completion time
3. There can be more than one critical path
4. Critical path has zero float

(a) All of the above (b) 1, 2 and 3
(c) 1, 2 and 4 (d) 1, 3 and 4

Q.70 Negative float means

(a) You are ahead of the schedule
(b) You are behind the schedule
(c) You are just on schedule
(d) It don't say anything of schedule

Q.71 Matching the following:

A. Total float 1. $T_L^j - T_E^i - t_{ij}$
B. Free float 2. $T_E^j - T_E^i - t_{ij}$
C. Independent float 3. $T_E^j - T_L^i - t_{ij}$

Codes:

	A	B	C
(a)	1	2	3
(b)	2	3	1
(c)	3	2	1
(d)	1	3	2

Q.72 A milestone
1. Represents completion of key deliverables
2. Is listed in project management plan and WBS dictionary
3. Is a scheduling Technique
4. Zero duration.

Which are correct.

(a) All of the above (b) 1, 2 and 4
(c) 2 and 3 (d) 3 and 4

Q.73 In Bottom - up method of estimating activity time/cost
1. Bottom up method is used
2. No past data present
3. WBS used

Which are correct

(a) All of the above (b) 1 and 2
(c) 2 only (d) 1 and 3

Q.74 Scheduling is done because
1. It gives start and end date for project.
2. It provides direction for project team to follow.
3. It makes project attractive.
4. To manage different activities timely.

(a) All of the above (b) 1, 2 and 3
(c) 1, 2 and 4 (d) 2 and 3

Q.75 Which among are schedule compression technique?
1. Crashing
2. Critical chain method
3. Resource leveling
4. Fast tracking

(a) All of the above (b) 1 and 2
(c) 1 and 4 (d) 1, 2 and 3

Q.76 In respect to crashing

(a) Total time reduces cost increases
(b) Total time reduces cost reduces
(c) Total time increases cost increases
(d) Total time increases cost decreases

Q.77 Which among the followings is graphical resource usage hours display method?

(a) Balance sheet (b) WBS
(c) Histogram (d) Risk Register

Q.78 Duration of milestone is

(a) Equal to first critical activity duration
(b) Zero duration
(c) equal to critical path duration
(d) equal to duration of project

Q.79 Which among the following is faster method of finding activity duration?

(a) Analogous method
(b) Bottom up method
(c) Can't say
(d) Both are very time consuming

Q.80 A dummy activity

(i) Consumes No time
(ii) Consumes No resources
(iii) Need not to follow Fulkerson rule

Which are correct?

(a) All of the above (b) (i) & (ii)
(c) (ii) & (iii) (d) (iii) only

Q.81 Which among the following are correct for PERT :

(i) It is event oriented
(ii) It is probabilistic in approach
(iii) Three time estimate is done

(a) All of the above (b) (i) & (ii)
(c) (ii) & (iii) (d) None of these

Q.82 Expected time in a PERT is given by

(a) $T_E = \dfrac{T_o + 4T_m + T_P}{6}$

(b) $T_E = \dfrac{T_P + 4T_o + T_m}{6}$

(c) $T_E = \dfrac{T_o + 4T_P + T_m}{6}$

(d) $T_E = \dfrac{T_o + 4T_m + T_P}{6}$

Q.83 Project management plan include

(i) Scope Baseline
(ii) Cost baseline
(iii) Project manager & Team members personnel details

(iv) Schedule plan

(a) All of the above (b) (i), (ii) & (iii)
(c) (i), (ii) & (iv) (d) (ii), (iii) & (iv)

Q.84 Which of the following are correct regarding kick off meeting :

(i) It is first meeting with project team and sponsor
(ii) Phase initiation and project initiation are necessary for kickoff meeting
(iii) Kickoff meeting occurs at start of every phase
(iv) Project team is evaluated at kickoff meeting.

(a) All of the above (b) (ii) & (iii)
(c) (i) & (ii) (d) (ii), (iii) & (iv)

Q.85 Find the standard deviation of the given PERT network.

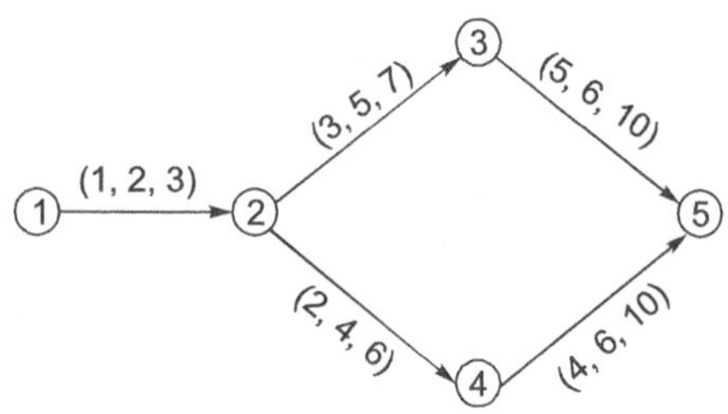

(a) $\frac{5}{2}$ (b) $\frac{\sqrt{5}}{2}$

(c) $\frac{5}{6}$ (d) $\frac{\sqrt{5}}{6}$

Q.86 Find the critical path in the given network.

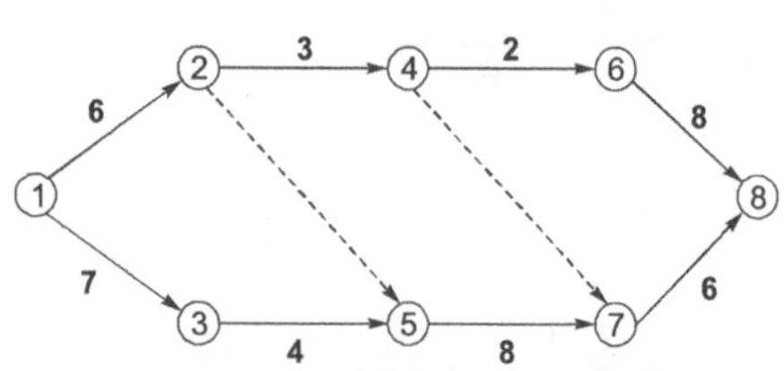

(a) (1)→(2)→(4)→(6)→(8)
(b) (1)→(2)→(4)→(7)→(8)
(c) (1)→(2)→(5)→(7)→(8)
(d) (1)→(3)→(5)→(7)→(8)

Q.87 In the given project network the duration of activities and cost slope are shown. Indirect cost per day is Rs. 200. The cost of the project is Rs. 30,000. Crash the project network to 11 days and find the revised project cost.

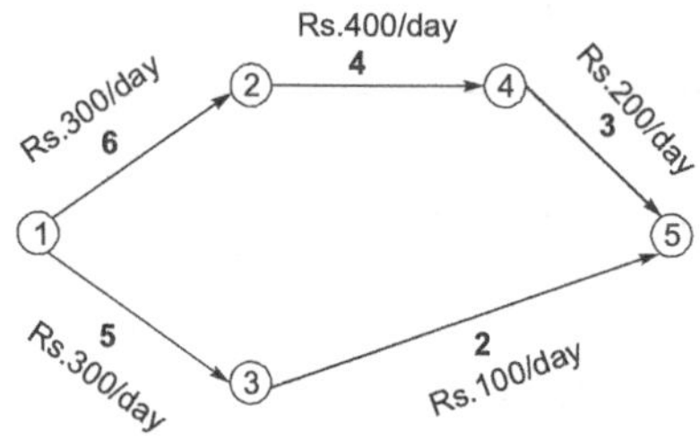

(a) 30,200 (b) 30,000
(c) 30,400 (d) 30800

Q.88 Find the free float of activity (2)→(4) in the given network

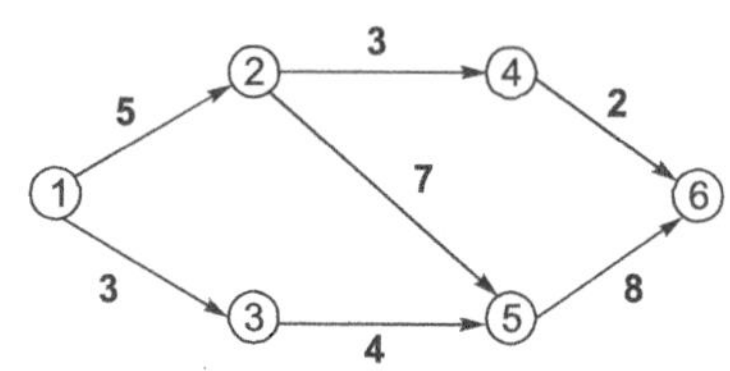

(a) 10 (b) 13
(c) 3 (d) 0

Q.89 In the given PDM network diagram find the slack of activity B. Activity duration are indicated.

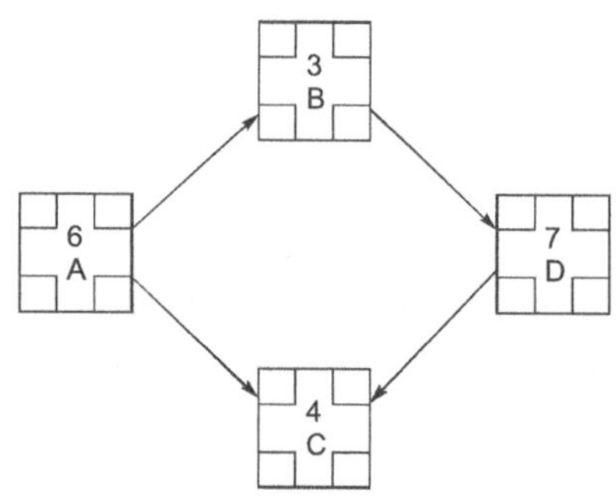

(a) 3 (b) 2
(c) 1 (d) 0

Q.90 For the given PERT network for what time duration will the completion probability be 99.87%? (Pert values are in days.)

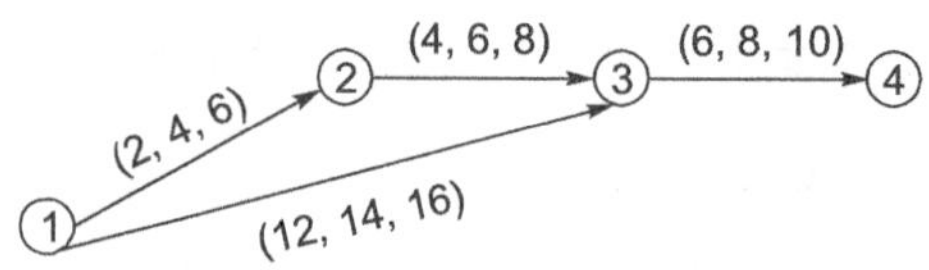

(a) 22 days
(b) 24 days

(c) $\left(22+2\sqrt{2}\right)$ days

(d) $\left(24+2\sqrt{2}\right)$ days

Q.91 The main difference between the two types (ADM, arrow diagramming method, and PDM, precedence diagramming method) of the critical path method (CPM) of scheduling is:

(a) Placement of the activity on the logic diagram line

(b) Arrow diagramming method (ADM) is a deterministic method, whereas the precedence diagramming method (PDM) is a probabilistic method.

(c) Precedence diagramming method (PDM) is a deterministic method, whereas the arrow diagramming method (ADM) is a probabilistic method

(d) Precedence diagramming method (PDM) is a more accurate method

Q.92 Which one of the following is captured in the Work Breakdown Structure (WBS)?

(a) The life cycle phases.

(b) The logical order of task

(c) The scope of the project

(d) Project costs

Q.93 Consider the following project network. Determine the critical path and the project duration.

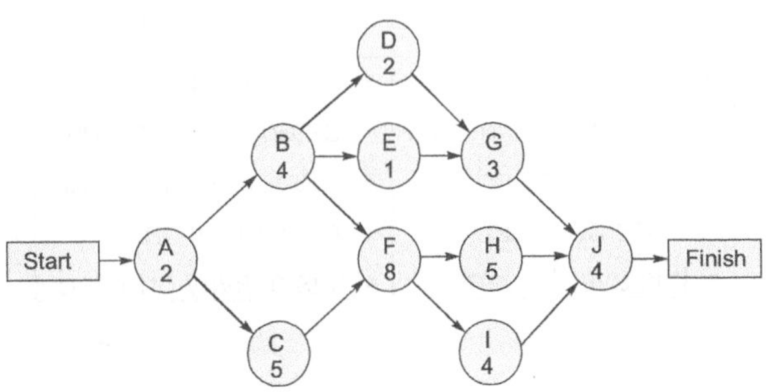

(a) The critical path is A-C-F-H-J with a completion time of 28 days

(b) The critical path is A-C-F-H-J with a completion time of 24 days

(c) The critical path is A-B-E-G-J with a completion time of 28 days

(d) The critical path is A-B-D-G-J with a completion time of 27 days

Q.94 The artificial activity which indicates that an activity following it, cannot be started unless the preceding activity is complete, is known as:

(a) Free float (b) Event

(c) Dummy (d) Constant

Q.95 The technique that allows us to redraw the activity network to more closely sequence project substasks to make the overall network sequence more efficient is known as :

(a) Lagging (b) Laddering

(c) Sequencing (d) Beta distribution

Answers

1.	(b)	2.	(d)	3.	(a)	4.	(b)	5.	(b)
6.	(a)	7.	(a)	8.	(a)	9.	(a)	10.	(b)
11.	(a)	12.	(c)	13.	(d)	14.	(a)	15.	(a)
16.	(b)	17.	(b)	18.	(c)	19.	(b)	20.	(a)
21.	(c)	22.	(b)	23.	(c)	24.	(b)	25.	(a)
26.	(a)	27.	(c)	28.	(a)	29.	(c)	30.	(c)
31.	(a)	32.	(a)	33.	(a)	34.	(d)	35.	(b)
36.	(b)	37.	(a)	38.	(c)	39.	(c)	40.	(a)
41.	(d)	42.	(d)	43.	(a)	44.	(c)	45.	(a)
46.	(a)	47.	(b)	48.	(b)	49.	(d)	50.	(b)
51.	(a)	52.	(c)	53.	(c)	54.	(b)	55.	(d)
56.	(d)	57.	(c)	58.	(b)	59.	(b)	60.	(c)
61.	(d)	62.	(a)	63.	(c)	64.	(a)	65.	(b)
66.	(c)	67.	(a)	68.	(a)	69.	(a)	70.	(b)
71.	(a)	72.	(a)	73.	(a)	74.	(c)	75.	(c)
76.	(a)	77.	(c)	78.	(b)	79.	(a)	80.	(b)
81.	(a)	82.	(a)	83.	(c)	84.	(c)	85.	(b)
86.	(d)	87.	(b)	88.	(d)	89.	(c)	90	(c)
91.	(a)	92.	(c)	93.	(b)	94.	(c)	95.	(b)

Explanations

1. (b)

A milestone is a significant point or event in the project. Milestones may be required by the project sponsor, customer, or other external factors for the completion of certain deliverables. They are similar to schedule activities, with the same structure and objectives, but they have zero duration as they represent a moment in time.

5. (b)
Work breakdown structure is a pictorial representation of major and minor objectives of a project. It is not a network. Instead it is a preliminary diagram of the way in which all the minor objectives go together to ensure the attainment of the major objective.

6. (a)
The bar chart does not show clearly the interdependencies between the various activities.

7. (a)

Paths	Duration (days)
$A-B-E-G$	$8 + 10 + 7 + 9 = 34$
$A-B-D-G$	$8 + 10 + 12 + 9 = 39$
$A-C-F-G$	$8 + 6 + 16 + 9 = 39$
$A-C-D-G$	$8 + 6 + 12 + 9 = 35$

Thus project duration is 39 days and there are two critical paths $A-B-D-G$ and $A-C-F-G$.

8. (a)

Expected time, $t_e = \frac{a+4m+b}{6}$

Standard deviation, $\sigma = \frac{b-a}{6}$

Activity	a	m	b	t_e	σ
A	4	6	8	6	0.67
B	7	10	13	10	1
C	6	15	18	14	2
D	8	8	8	8	0
E	4	5	6	5	0.33
F	5	5	5	5	0

The critical path will be $A-B-C-F$ and the project duration will be 35.

∴ Standard deviation of the project,

$$\sigma = \sqrt{(0.67)^2+(1)^2+(2)^2+0}$$

$$= 2.33$$

Range of project duration

$$= T_E - 3\sigma \text{ to } T_E + 3\sigma$$

$$= 35 - 3 \times 2.33 \text{ to } 35 + 3 \times 2.33$$

$$= 28 \text{ to } 42$$

Proportional variation

$$= \frac{\text{Standard deviation}}{\text{Project duration}}$$

$$= \frac{2.33}{35} = \frac{1}{15}$$

10. (b)
Rules for drawing Network Diagram:
(i) Each activity is represented by one and only one arrow in the network.
(ii) No two activity can be identified by the same end events.
(iii) In order to ensure the correct precedence relationship in the arrow diagram, following question must be checked whenever any activity is added to network.
(a) What activity must be completed immediately before this activity can start?
(b) What activity must follow this activity?
(c) What activity must occur simultaneously with this activity?

11. (a)
Project management has evolved two analytical techniques for planning, scheduling and controlling of projects. These are the Critical Path Method (CPM) and the project evaluation and Review Technique (PERT).
CPM – Activity Oriented
PERT – Event Oriented
CPM – Activity Oriented

13. (d)

PERT	CPM
• It is based on probabilistic model • Time is important consideration. • It is event oriented	• It is based on deterministic model. • Money is important consideration. • It is activity oriented.

14. (a)
Milestones are singular points in time, such as the start or completion of a significant activity or group of activities.

15. (a)
Total float or slack is computed by subtracting the early start date from the late start date, or 19 – 10 = 9. To compute the early finish date given a duration of 4, we would start counting the activity on the morning of the 10th; therefore, the activity would be completed at the end of day

13, not 14 (10, 11, 12, 13). If we started the activity on its late start date on the morning of the 19th, we would finish at the end of day 22, not 25. Insufficient information is provided to determine whether this activity can be completed in 2 days if the resources are double.

16. (b)
Free float is defined as the amount of time an activity can be delayed without delaying the early start of any immediately succeeding activities.

17. (b)
First, compute the standard deviation:

$$\sigma = \frac{P-O}{6} \text{ or } \frac{36-6}{6} = 5 \text{ days}$$

Next, compute PERT expected time:

$$\frac{P+4(ML)+O}{6} \text{ or } \frac{36+4(21)+6}{6} = 21 \text{ days}$$

Finally, determine range of outcomes using 1σ: 21 – 5 = 16 days, and 21+5 = 26 days

Simply defined, 1σ is the amount on either side of the mean of a normal distribution that will contain approximately 68.26 percent of the population.

18. (c)
In analyzing a project the first step is always Work Breakdown Structure (WBS). From WBS, network can be prepared. Thereafter scheduling with resource allocation follows. Finally time - cost study is carried out.

21. (c)
Try to find those activities delay which causes delay in whole project.

26. (a)
Three time estimates for each activity are required in PERT network. PERT is built up of event oriented diagram.

31. (a)
A work package is the lowest or smallest unit of work division in a project or WBS. The work package can be scheduled, cost estimated, monitored, and controlled.

44. (c)
In analyzing a project, the first step is always Work Breakdown Structure (WBS). From WBS, network can be prepared. Thereafter scheduling with resource allocation follows. Finally time - cost study is carried out.

45. (a)
Door – window shutters can be fixed only after door – window frames.

Wall plastering cannot be done prior to conceal wiring.

Therefore correct sequence is 4–3–2–5–1. i.e. (Fixing door – window frames) – (Conceal wiring) – (Wall Plastering) – (Fixing door – window shutters) – (Flooring).

47. (b)
Work breakdown structure is a pictorial representation of major and minor objectives of a project. It is not a network. Instead it is a preliminary diagram of the way in which all the minor objectives go together to ensure the attainment of the major objective.

49. (d)
The delay which can be performed without delaying succeeding activities and affects only preceding activities is called free float. Free float (F_F) is given by

$$F_F = \left(T_E^j - T_E^i\right) - t_{ij}$$

Total float is the time span by which starting or finishing of an activity can be delayed without delaying the completion of the project. Total float (F_T) is given by

$$F_T = \left(T_L^j - T_E^i\right) - t_{ij}$$

Independent float is the minimum excess available time which exists without affecting any of succeeding or preceding activities. It is given by

$$F_{ID} = \left(T_E^j - T_L^i\right) - t_{ij}$$

Interfering float (F_{INT}) is the difference of total and free float.

$$F_{INT} = F_T - F_F$$

50. (b)
Slack denotes the flexibility range within which an event can occur.
Slack of an event = (LOT – EOT) of the event. In the CPM network the minimum value of slack is zero and the zero slack event is called the critical event.

51. (a)
In PERT time estimate follows beta distribution. In CPM it follows normal distribution.

56. (d)
Schedule compression techniques are used when a project gets behind the schedule.

61. (d)
Simulations are typically performed using Monte Carlo in which a project model is computed many times with the input values chosen at random for each iteration from the probability distribution of these variables. Monte Carlo analysis supports various statistical distributions (normal, triangular, beta, uniform, etc. used in estimating budgets, schedules, and resource allocations.

63. (c)
All projects don't require these knowledge areas uniformly. There might be situation where HR is on higher priority than any other.

66. (c)
Performance measurement baseline is earlier agreed scope and conditions. They can't be changed be anyone except sponsor.

70. (b)

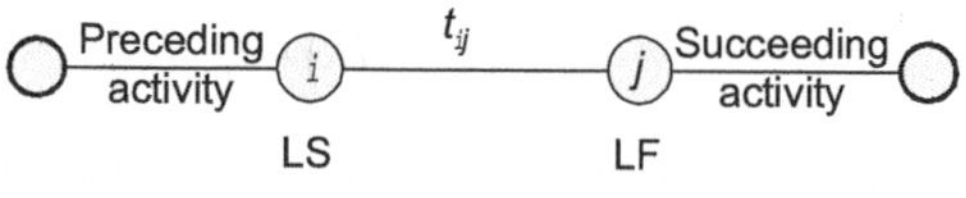

Where, ES → Early start
EF → Early finish
LS → Late start
LF → Late finish

74. (c)
Scheduling tools the interconnection between different activities on time basis. It has nothing to do with attractiveness.

78. (b)
Milestone has zero duration.

83. (c)
Project management plan don't include personal details of any employee.

84. (c)
Phase Initiation + Project initiation $\Longrightarrow$ Kick off meeting

85. (b)
Critical path is (1) → (2) → (3) → (5)

$$t_{e_1} = 2$$
$$t_{e_2} = 5$$
$$t_{e_3} = \frac{5+24+10}{6} = \frac{39}{6} = 6\frac{1}{2}$$
$$t_e = 13\frac{1}{2}$$
$$\sigma_1 = \frac{1}{3};\ \sigma_2 = \frac{2}{3};\ \sigma_3 = \frac{5}{6}$$
$$\sigma = \sqrt{\frac{1}{9}+\frac{4}{9}+\frac{25}{36}}$$
$$= \sqrt{\frac{4+16+25}{36}}$$
$$= \sqrt{\frac{45}{36}} = \frac{\sqrt{45}}{6} = \frac{\sqrt{5}}{2}$$

88. (d)

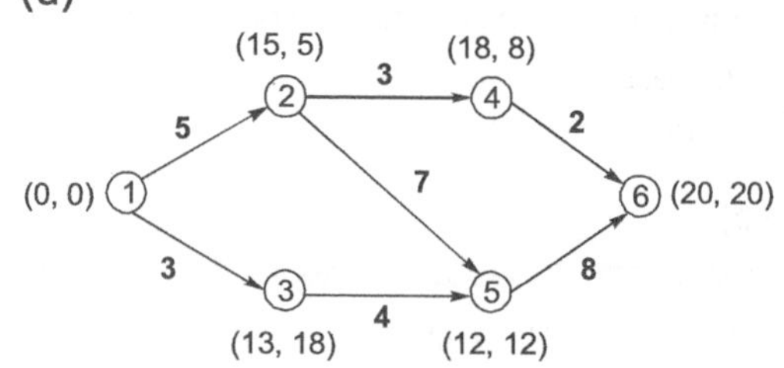

89. (c)

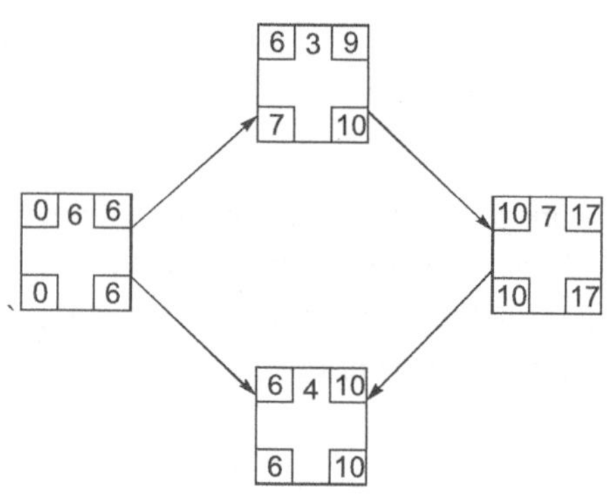

90. (c)

Critical path is (1) → (3) → (4)

$$t_{e_1} = 14$$

$$t_{e_2} = 8$$

$$t_e = 14 + 8 = 22$$

Also, $\sigma_1 = \frac{2}{3}; \sigma_2 = \frac{2}{3}$

$$\therefore \quad \sigma = \sqrt{\sigma_1^2 + \sigma_2^3}$$

$$= \sqrt{\frac{4}{9} + \frac{4}{9}} = \frac{2\sqrt{2}}{3}$$

For completion probability to be 99.87%

$$\text{Time needed} = t_e + 3\sigma$$

$$= 22 + 3 \times \frac{2\sqrt{2}}{3}$$

$$= 22 + 2\sqrt{2}$$

92. (c)

The work breakdown structure approach allows us to visually see the work that is needed in order to complete a project. The bottom line is by using a work breakdown structure it reduces the number of surprises and improves the ability to better estimate future projects. The work breakdown structure can be used to effectively decompose the project scope, to improve estimating, to better control the project execution and to more accurately verify project completion. In addition, using a work breakdown structure approach summarizes project information to improve the opportunity for use of historical information, which, can aid in both speed and accuracy of future projects. The work breakdown structure is a reportable process that can be used as a template for future projects.

93. (b)

A-C-F-H-J with a completion time of 27 days as calculated by adding the time required to traverse the path.

94. (c)

Dummy activities are a useful tool to implement when the specific logical relationship between two particular activities on the arrow diagramming method cannot specifically be linked or conceptualized through simple use of arrows going from one activity to another. The time taken for the dummy activities is zero.

■■■■

6 EXECUTION PROCESS: Project Monitoring and Control

6.1 PROJECT MONITORING AND CONTROL

According to the Project Management Body of Knowledge (PMBOK), "the Monitoring and Control Process Group consists of those processes performed to observe project execution so that potential problems can be identified in a timely manner and corrective action can be taken, when necessary, to control the execution of the project."

Project Monitoring and Control activities take place in parallel with Project Execution Process Group activities so that, while the project work is being executed, the project is being monitored and controlled by implementing the appropriate level of oversight and corrective action.

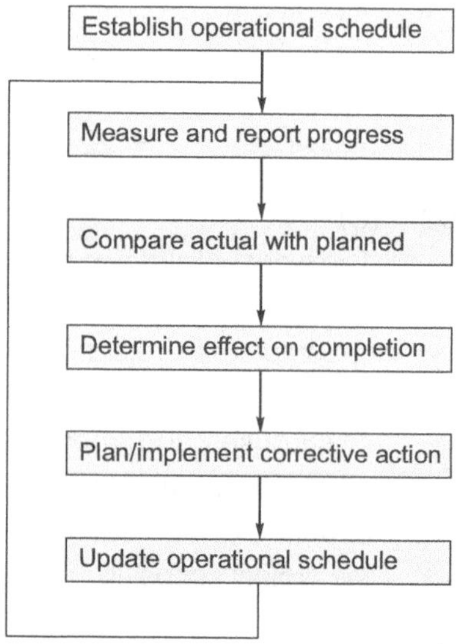

The project is observed and measured regularly against the project plan to ensure that the project is within acceptable variances of cost, schedule and scope, and that risks and issues are continually monitored and corrected as needed.

The main purpose of monitoring and controlling activities is to be proactive in finding issues ahead of time and taking corrective action. Corrective action can require revisiting Planning Process Group and updating the Project Management Plan as needed with the ultimate goal of bringing the project back in line with project objectives and constraints and improving future execution to avoid repeating the same issues.

6.2 PROJECT CONTROL SYSTEM

- Monitoring is the collection, recording and reporting of the project data and information whereas control uses the monitored data and information to bring actual performance into agreement with the plan. Although the data gathered is primarily used for control but can also be used for auditing and improving quality.
- Control system is an arrangement that offers the project manager with details about deviations of the project from what was planned and also recommends corrective actions needed for rectifying the deviations.
- Control systems are designed to monitor three major factors, viz., cost, time and quality parameters. In other worlds, control systems help in identifying cost over-run situations, time over-run situations and deviations from quality parameters.
- Since projects are dynamic in nature, project control systems are to be more sophisticated than the control systems meant for functional departments. The sophistication of the control system depends on the nature and complexity of projects and also on the ability of the project participants to administer it.

Requirements of a Good Control System:

- It must be easy to understand.
- It must be easy to extract data/information.
- It must act as an indicator for pointing out deviations.
- It must be reduced to the form of tables/graphs/charts so that it will offer a visual display of the happenings.
- The control system should report deviations (of time, cost and quality) from the plan on a timely basis and must have the capacity to anticipate or predict deviations so that timely action can be taken to correct the deviations.
- The control system is to be designed by the active participation of all the major executives of the project team so that the system can satisfy the actual requirements.

6.3 TOOLS USED FOR PROJECT CONTROL

- Work breakdown structure
- Bar chart/Mile stone chart
- Line of Balance (LOB) chart
- Critical ratio
- Earned Value analysis
- Network plans (Precedence diagrams, CPM/PERT networks)
- Graphical Evaluation Review Technique (GERT)
- Network simulation
- Out sourcing method
- Monthly status reports
- Review meetings
- S-curve

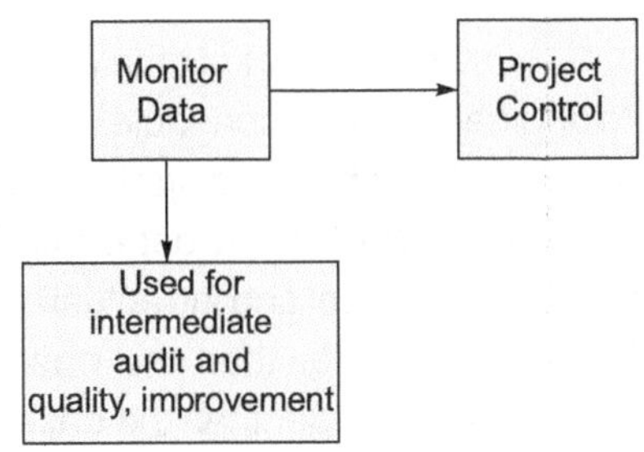

(i) **S-Curve :** They allow quickly identify Project growth, slippage and other problems directly by simply observing the *s*-curve. Determine growth of project using *s*-curve. If the target S-curve finishes above base line S-curve; the project has grown and if it finishes below base line S-curve, it has contracted.

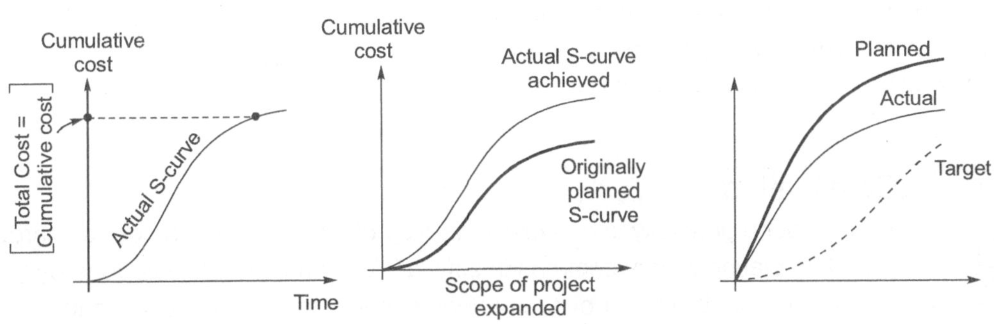

If the base line changes, scope also changes.

6.4 DETERMINING GROWTH

Comparison of the Baseline and Target S-curves quickly reveals if the project has grown (Target S-curve finishes above Baseline S-curve) or contracted (Target S-curve finishes below Baseline S-curve) in scope. A change in the project's scopes implies a re-allocation of resources (increase or decrease), and the very possible requirement to raise contract variations. If the resources are fixed, then the duration of the project will increase (finish later) or decrease (finish earlier), possibly leading to the need to submit an extension of time claim.

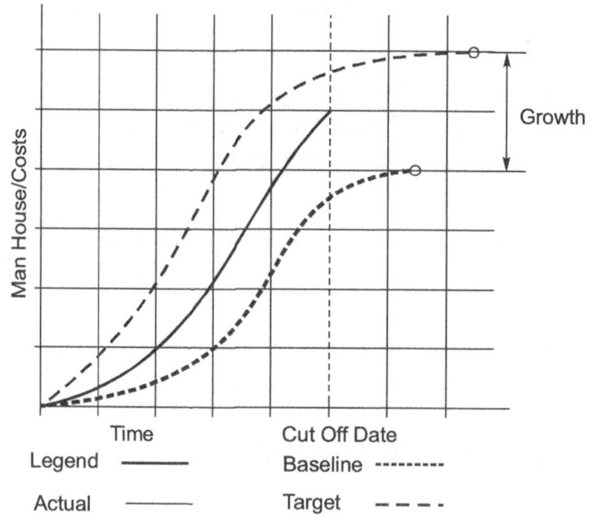

6.5 2-DETERMINING SLIPPAGE

"The amount of time a task has been delayed from its original baseline schedule. The slippage is the difference between the scheduled start or finish date for a task and the baseline start or finish date. Slippage can occur when a baseline plan is set and the actual dates subsequently entered for tasks are later than the baseline dates or the actual durations are longer than the baseline schedule durations".

Comparison of the Baseline S-curve and Target S-curve quickly reveals any project slippage (i.e. the Target S-curve finishes to the right of the Baseline S-curve). Additional resources will need to be allocated or additional hours worked in order to eliminate (or at least reduce) the slippage. An extension of time claim may need to be submitted if the slippage cannot be eliminated or reduced to an acceptable level.

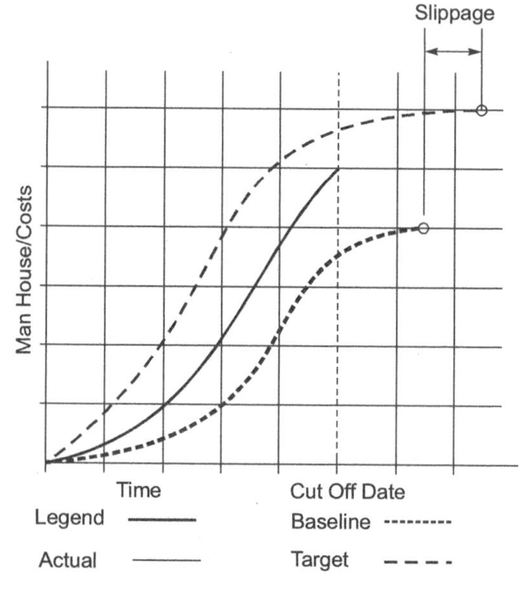

NOTE

Gold Plating

- It is done generally to be appreciated by customer, no extra cost is charged for it.
- Gold plating is very common in software programming and is done by team members to show their abilities, or by the project manager to make the client happy, though customer may or may not like these changes.

NOTE: 1. Main aim here is to monitor & control various processes of the project.
2. It is an umbrella process which is present in other groups also.

6.6 EARNED VALUE ANALYSIS (EVA)

Effective control over a project can be exercised only by performing periodic and systematic performance analysis. Modern, method of performance analysis (also called Earned Value Analysis) provides an analytical framework for project control by estimating factors like cost variance, schedule variance, cost performance index, schedule performance index, estimated cost performance index etc.,

Arriving at the above parameters for a project needs the understanding of the following terms.

(i) Cost Variance (CV)

It is difference between the actual expenditure made in the project till the date of review to the value of work accomplished for the expenses incurred.

$$CV = BCWP - ACWP \qquad \text{BCWP also called earned value EV}$$
$$= EV - AC \qquad \text{ACWP also called actual cost AC}$$

(ii) Cost Performance Index

$$\text{CP Index} = BCWP/ACWP = \frac{EV}{AC}$$

[CP Index of less than 1.00 indicates that the project is spending more than the scheduled cost.]

(iii) Schedule Variance (SV)

It is the difference between the value of work that the has actually been completed and the value of the work schedule to have been completed.

$$SV = BCWP - BCWS \qquad \text{BCWS also called planned value PV}$$
$$= EV - PV$$

[A negative SV indicates that the project is behind the schedule.]

The above formula gives the schedule variance in cost terms.

(iv) Schedule Performance Index

This parameter gives the schedule variance in the form of an index.

$$\text{SP Index} = BCWP/BCWS$$
$$= \frac{EV}{PV}$$

[SP Index of less than 1.00 indicates that the* progress of work performed is less than the schedule.]

(v) Time Variance (TV)

The Time Variance gives the time lag, at the date of review, by indicating the time equivalent of the value of work which has to be completed to stay on schedule.

$$TV = (\text{Date of review}) - (\text{Date on which } BCWP = BCWS)$$

(vi) Estimated Cost Performance Index

$$\text{ECP Index} = BCTW/(ACWP + ACC)$$

[ECP Index of less than 1.00 indicates the amount that will be spent on the project will be more man the budgeted amount.]

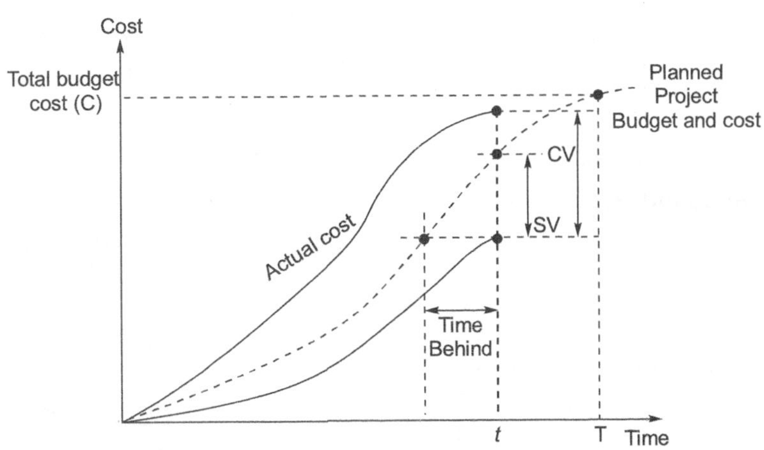

t – Time of earned value analysis
C – Total Planned budget cost
SV – Schedule Variance
T – Planned time of completion
CV – Cost Variance

Cost Variance (CV)	CV = BCWP – ACWP	CV = EV – AC
Cost Performance Index	$CPI = \frac{BCWP}{ACWP}$	$CPI = \frac{EV}{AC}$
Schedule Variance (SV)	SV = BCWP – BCWS	SV = EV – PV
Schedule Performance Index	SPI = BCWP/BCWS	$SPI = \frac{EV}{AV}$
Estimated Cost Performance Index	$ECPI = \frac{BCTW}{(ACWP + ACC)}$	$ECPI = \frac{BAC}{EAC}$

EXAMPLE 6.1

XYZ Company has bagged a fixed cost contract for the supply, installation, testing and commissioning of 200 computers of same specifications at a cost of Rs. 600 lakhs. The company had estimated that it could supply, install, test and commission 10 computers per day so that the entire work would be completed in 20 days time. The project status was reviewed after the completion of 16 days. It was noted at the time of review that only 120 computers have been installed and the cost incurred was Rs. 380 lakhs. It was estimated at the time of the review that a sum of Rs. 260 lakhs would be required for completing the pending work, viz., installation of the remaining 80 computers.

Make a performance analysis by arriving at the various connected parameter

Solution:

BCWS:

Work scheduled per day : 10 computer installation

Work scheduled for 16 days (16 x 10) : 160

Budgeted cost of one installation (600/200) : Rs. 3 lakhs

BCWS

BCWP:

After 16 days, only 120 computers have been installed.

Hence, BCWP : 120×3

: Rs. 360 lakhs

Cost Variance:

CV : BCWP – ACWP

: $(120 \times 3) - 380$

: Rs (–) 20 lakhs*

(*The negative figure indicates that the actual cost is more than the budgeted cost)

Cost Performance Index:

CP Index : BCWP/ACWP

: 360/380

: 0.947

Scheduled Variance (in cost terms):

SV : BCWP – BCWS

: 360 – 480

: Rs. (–) 120 lakhs

Schedule Performance Index:

SP Index : BCWP/BCWS
: 360/480
: 0.75

Time Variance:

No. of installations that ought to have been completed after 16 days (16 x 10)	: 160
No. of installations actually completed after 16 days	: 120
Short fall	: 40

TV is the time equivalent of work which has to be completed to stay in schedule, i.e., the number of days required to make good the short fall of 40, the rate of installation per day as planned being 10.

TV : 40/10
: 4 days

Estimated Cost Performance Index:

ECP Index : BCTW/(ACWP + ACC)
: 600/(300 + 260)
: 0.9375

6.7 CRITICAL RATIO

- A critical ratio indicates a manager that when a task or process is becoming unacceptable, typically when the ratio drops below one then project is unacceptable.
- The calculation of the critical ratio for project tasks is the product of a schedule *ratio* and a *cost ratio.*
- The schedule ratio is actual progress divided by scheduled progress, as measured by some common standard such as earned value: EV/PV. Clearly, ratios greater than one are desirable.
- The cost ratio is budgeted cost divided by actual cost, or if earned value data is available, EV/AC. Again, values greater than one are most desirable.
- Taking the product of these two ratios thus gives us an overall measure that includes performance, cost, and schedule.

  CR = (actual progress/scheduled progress) × (budgeted cost/actual cost)

6.8 LINE OF BALANCE (LOB)

- Line of balance is a planning technique for repetitive work. In this technique, the required resources for each stage or operation are found out so that, the subsequent stages of the activities are not interfered with and the target output can be achieved. The technique can be applied in repetitive works involving construction projects such as mass housing developments, high rise buildings, tunnels etc.
- In a repetitive fashion of construction projects, cost and time efficiencies can be achieved by providing standardisation of design for different units of work. It can be achieved by balancing the crews with each other and with constraining resources. By such planning, we can achieve a continuity in the placement of all repetitive elements, thus maximising the productivity of labour and equipment.
- The technique of LOB can be used for assembling, selecting, interpreting and presenting in graphical form, the essential factors involved in construction from initial stage to the completion of construction against a background of time. This technique is highly effective in determining areas of weakness and focusing on items requiring immediate attention.

- Application of LOB in any construction situation will involve identification of some stations or stages in the whole sequence of operations and scheduling and monitoring process at these stations, so that the whole project progresses as per agreed schedule. The mile stone concept becomes handy in this context. The technique may use these milestones keeping in view the balancing requirement of the other milestones and the project as a whole.
- For mass construction, LOB may show the volume of work required to be completed at important stages to meet a contractor's commitment to the owner. When LOB is used, the following two things are likely to be achieved:
  (i) Work will be completed as per commitment made.
  (Ii) At each important stage, work will be completed only when needed. This means resources will not be spent until they are required.

6.9 GRAPHICAL EVALUATION AND REVIEW TECHNIQUE (GERT)

- Precedence networks and decision CPM networks are 'Deterministic' in nature. That is, they do not involve 'Probabilistic' elements where as GERT is based on probabilistic approach.
- A number of researchers have proposed and expanded more general networking systems in order to expand the modelling capabilities of PERT and CPM. Most of these make use of what are called 'Probabilistic nodes'. These differ considerably from ordinary project network nodes (i.e.; events). In the latter case, every node is eventually reached or released when all the jobs leading into the node have been completed. Similarly, all jobs proceeding from the node may start after the node is released and all of them must eventually be completed.
- But probabilistic networks show that certain nodes may be released when one or more (but not necessarily all) jobs leading into them are completed. These networks also show that there is a possibility of probabilistic exits from the node, in which case only one of several proceeding jobs is actually taken. Thus, jobs themselves may be probabilistic, occurring with some probability, which may be less than 1.

6.10 NETWORK SIMULATION

- PERT assumes three probable time estimates for each activity and assumes a specific distribution for arriving at the expected time of completion of each activity. However, in actual practice, the distribution of time estimates is not fixed, but will vary according to the real life situations.
- Network Simulation models permit assuming any activity-time distribution. For a given set of activity-time distributions, the model simulates 'Project duration distribution'. The project duration distribution is the output of Network Simulation models while the activity-time distributions of the different activities of the project form the input to the Net Work Simulation Model.
- Since a Network Simulation model can accept any type of distribution for the activity-time, it resembles real life situations and permits assignment of any realistic activity-time distribution so that the project completion time can be more realistically assessed.

6.11 RESOURCE ALLOCATION

- A project manager comes across resource constraints. Either there may be delay in arrival of material resources or there may be shortfall of human resource or some time on some day human resource may be excess.

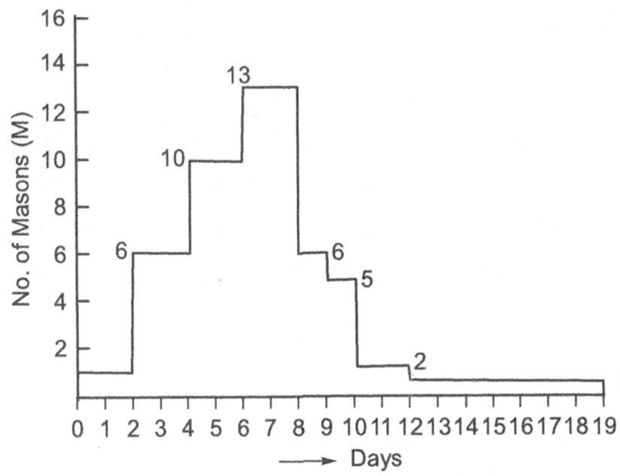

Fig. *A Typical Histogram for Resource Allocation*

- The job of the project manager is to plan and allocate the resources for the different activities such that the resource utilisation is optimised.
- The resource is a physical variable which is required for completion of an activity. It can be material, man power, machinery, space, money or time resource.
- Resource allocation means the decision to allocate the resources for each activity. Due to limited resources or limited project duration, it is required to allocate the resources such that more or less uniform demand is generated throughout the project duration.
- The diagram which shows the variation of the requirements with the time is called 'resource usage profile' or 'resource histogram'.

  Resource allocation may be achieved by the following two processes:

  A. Resource levelling

  B. Resource Smoothing

6.12 RESOURCE LEVELLING

- Here the resources are considered unlimited but the project duration is unchanged or fixed.
- Resource levelling aims to reduce the peak requirements and smooth output period to period assignment without prolonging overall project duration.
- There are situations demanding that the project should be completed by a specified due date. The due-date for project completion is decided by the management for various reasons.
- For example, a canal lining project might be required to be completed before the monsoon sets on; a school building might be required to be completed before the school reopens after vacation. Under such situations, project completion time is the constraint *i.e.*, the project is to be completed at any cost by the due-date.
- A project manager often comes across mis-match between the availability of resources and the requirement of resources. This means that there are surplus resources available on some days and there is shortage of resources on some other days. For example, in construction industry, the requirement of personnel for executing construction work varies very much from day to day. It will be uneconomical to permanently employ personnel to satisfy the peak time requirements since there will be under utilization of personnel on many days. Hence, only less number of personnel than the peak time requirement is employed by the management with the hope that any shortfall in personnel requirement during peak time can be handled by hiring personnel on casual basis. Since the personnel requirement

varies from day to day, the project manager should plan properly in such a way that optimum utilization of available personnel is achieved.

- Thus, the objective of the project manager in such a situation is to level as far as possible the demand for resources throughout the project execution time, keeping in view the constraint that the "specified" project completion time should not be exceeded. The "specified" project completion time may be the minimum possible time by which the project can be completed as ascertained by identifying the critical path or even a shorter duration than this as decided by the management to meet other requirements.

Ideal Resource Profile

- While project scheduling is done it is assumed that human resources will be available as and when required but it is rarely possible.
- The employment and utilization is mostly stable and more at a constant rate. The ideal profile of resource utilization is as follows:

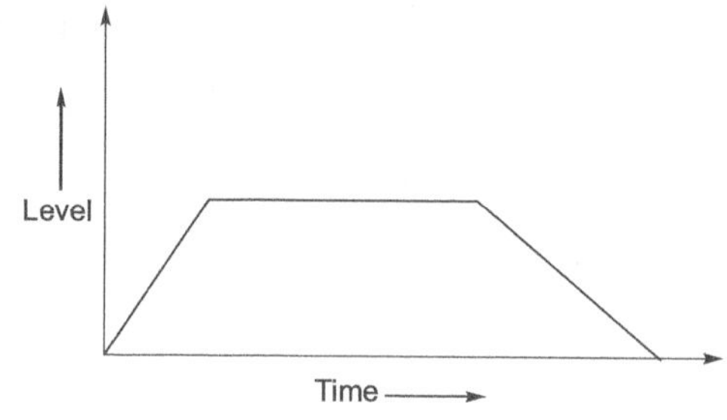

Since the engagement and release of manpower has to be in concrete steps hence the modified profile may be as below:

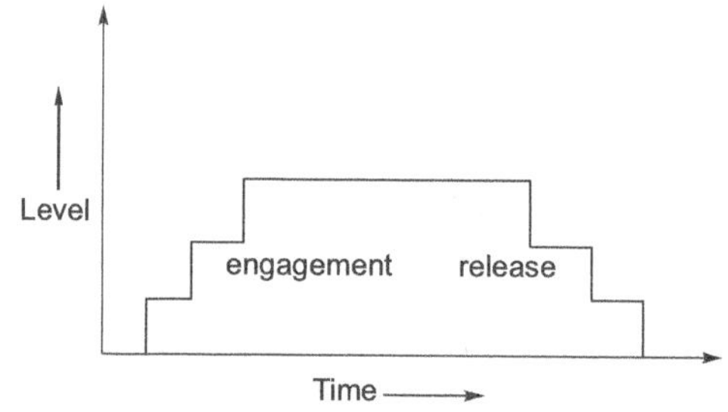

6.13 RESOURCE SMOOTHING

- Here the resources are considered limited, but the project duration may be changed. Activities are rescheduled to cut down the peak requirement of resources so that it does not cross the limit of resources. Available resources should never be less than the maximum quantity required for any activity of project. Firstly, available floats are used then if needed duration of some activities is increased or decreased as per the resource requirement.

NOTE

Conflict in Definitions
Resource levelling and resource smoothening defined in this book are as per definitions/concepts explained in Foreign author books such or Roury Bourke, Harald Kerzer and some Indian authors also have followed the same concept like M.R. Gopalan, K. Nagranjan, Prasanna Chandra, NPTEL Lectures etc. but some of the Indian authors have given just opposite explanation of these terms which include B.C. Puramia, R. Seetharaman and U.K. Shrivastava.

Previous ESE Prelims Question

Q.1 Crashing is
(a) Abandoning the project
(b) Completing the project with all possible haste
(c) Reduction of duration for a few of the activities
(d) Reducing the cost of the project with all needful modifications

[ESE-2018]

Ans. (c)

Q.2 **Statement (I)** : Social involvement discourages additional government regulation and intervention.
Statement (II) : Social involvement can create a weakened international balance of payments situation.
(a) Both Statement (I) and Statement (II) are individually true and Statement (II) is the correct explanation of Statement (I)
(b) Both Statement (I) and Statement (II) are individually true, but Statement (II) is not the correct explanation of Statement (I)
(c) Statement (I) is true, but Statement (II) is false
(d) Statement (I) is false, but Statement (II) is true

[ESE : 2022]

Ans. (c)

Q.3 Which of the following factors are included in product realization process?
1. Marketing functions to assess customer requirements
2. Documentation of the design
3. Legal requirements

Select the correct answer using the code given below:
(a) 1 and 2 only (b) 2 and 3 only
(c) 1 and 3 only (d) 1, 2 and 3

[ESE : 2022]

Ans. (d)

Q.4 **Statement (I):** The machine shop produces parts machined from stock material and finishes castings, forgings, etc., requiring machined surfaces.
Statement (II): In machine shops, machining operations remove metal, either to make a smoother and more accurate surface, as by planning, facing, milling, etc. or to produce a surface previously existing, as by drilling, punching, etc.
(a) Both Statement (I) and Statement (II) are individually true and Statement (II) is the correct explanation of Statement (I)
(b) Both Statement (I) and Statement (II) are individually true, but Statement (II) is not the correct explanation of Statement (I)
(c) Statement (I) is true, but Statement (II) is false
(d) Statement (I) is false, but Statement (II) is true

[ESE : 2023]

Ans. (c)

Objective Brain Teasers

Q.1 Consider the following statements regarding updating of the network:
1. For short duration project, updating is done frequently
2. For large duration project, frequency of updating is decreased as the project is nearing completion
3. Updating is caused by over estimated or under estimated times of activities.
4. The outbreak of natural calamity necessitates updating

Which of these statements are correct?
(a) 1, 2 and 3 (b) 2, 3 and 4
(c) 1, 3 and 4 (d) 1, 2 and 4

Q.2 If actual cost is more than the estimated cost then
(a) cost performance index CPI < 1
(b) CPI > 1

(c) CPI = 1
(d) Can't be predicted

Q.3 Your customer wants the project to be completed six months earlier than plan. You believe you can meet this target by overlapping project activities. The approach you plan to use is known as:
(a) Critical chain (b) Fast tracking
(c) Leveling (d) Crashing

Q.4 What is the process of incorporating changes and rescheduling or replanning called?
(a) Resource allocation
(b) Resource smoothing
(c) Resource levelling
(d) Updating

Q.5 In precedence diagram the relation is from completion of one activity to other activity is
(a) Start to start (b) Start to finish
(c) Finish to start (d) Finish to finish

Q.6 In the cost optimization procedure the cost slope for each activity can be estimated by the formula

(a) $\dfrac{\text{crash cost} - \text{normal cost}}{\text{crash time} - \text{normal time}}$

(b) $\dfrac{\text{crash time} - \text{normal cost}}{\text{crash cost} - \text{normal cost}}$

(c) $\dfrac{\text{normal time} - \text{normal cost}}{\text{crash cost} - \text{normal cost}}$

(d) $\dfrac{\text{crash cost} - \text{normal cost}}{\text{normal time} - \text{crash time}}$

Q.7 The normal duration and normal cost of an activity are 10 days and Rs. 350/- respectively. The cost slope is Rs. 75/- per day. If the crash duration is 8 days, then what is the crash cost of the activity?
(a) Rs. 400/- (b) Rs. 500/-
(c) Rs. 600/- (d) None of these

Q.8 In the time-cost optimization of a project, the project can be crashed by expending
(a) all activities on the critical path
(b) critical activities having minimum cost slope
(c) activities on subcritical path
(d) critical activities having maximum cost slope

Q.9 Match **List-I** (Cost) with **List-II** (Features) and select the correct answer using the codes given below the lists:

List-I
A. Optimal cost
B. Overhead cost
C. Direct cost
D. Indirect cost

List-II
1. Activity related
2. Developed by crashing process
3. Project-related
4. Contained in or contributing exclusively to the related product

Codes:

	A	B	C	D
(a)	4	3	2	1
(b)	2	1	4	3
(c)	4	1	2	3
(d)	2	3	4	1

Q.10 In big construction projects, the % cost of construction equipment in terms of the total project cost is about
(a) 10-20% (b) 15-30%
(c) 45-50% (d) 5-10%

Q.11 The objectives of cost control is
(a) To locate the areas of inefficient functioning
(b) Basis for preparing estimates
(c) Estimation of profit or loss
(d) All of these

Q.12 The stage of the work at which cost control is not affected is at
(a) Planning stage
(b) Designing stage
(c) Construction stage
(d) None of these

Q.13 The time beyond which the direct cost will not be reduced with the increase in time is
(a) Normal time (b) Crash time
(c) Optimistic time (d) Optimum time

Q.14 The direct cost of a project with respect to normal time is
(a) Minimum (b) Maximum
(c) Zero (d) Infinite

Q.15 The reduction in project time normally results in
(a) Increasing the direct cost and decreasing the indirect cost

(b) Decreasing the direct cost and increasing the indirect cost
(c) Increasing both the direct cost and indirect cost
(d) Decreasing both the direct cost and indirect cost

Q.16 Economic saving of project time would result by crashing
(a) Cheapest critical activity
(b) Cheapest non-critical activity
(c) Costliest critical activity
(d) Costliest non-critical activity

Q.17 In construction project, the cost-slope of an activity is an indication of
(a) extra-time needed
(b) extra cost needed
(c) reduction of duration of critical activity
(d) crashing of an activity

Q.18 Which of the following are examples of indirect costs?
(a) Salaries of corporate executives
(b) Salaries of full-time project staff
(c) Overhead costs
(d) Both (a) and (c)

Q.19 Which of the following is not a commonly used technique for schedule compression?
(a) Resource reduction
(b) Reducing scope
(c) Fast-tracking activities
(d) Use of overtime

Q.20 Overall cost estimates must be allocated to individual activities to establish the cost performance baseline. In an ideal situation, a project manager would prefer to prepare estimates:
(a) Before the budget is complete
(b) After the budget is approved by management
(c) Using a parametric estimating technique and model specific for that project type
(d) Using a bottom-up estimating technique

Q.21 The key to effective cost control is:
(a) Using earned value to forecast project status
(b) Focusing on projected expenditures and actively networking with key stakeholders to ensure funds will be available as requested
(c) Informing stakeholders of the project's cost status
(d) Managing the approved cost baseline and any changes to it

Q.22 What is the significant purpose of monitoring a project through its implementation phase?
(a) To fix responsibility for delays
(b) To retail the project with control over cost over-run
(c) To retail the project with minimum time over-run
(d) To retail the project with optimal time and cost over-run

Q.23 Which of the following is not an input to execution process?
(a) Enterprise environmental factor
(b) Organisational process assets
(c) Project management plan
(d) Accepted deliverables

Q.24 Which of the following are output of project execution process?
1. Deliverables
2. Project documents update
3. Project change requests
4. Project management plan

(a) All of the above (b) 1, 2 and 3
(c) 2, 3 and 4 (d) 2 and 3 only

Q.25 Which is not a tool for the project execution process?
1. Expert Judgement
2. Project management information system
3. Meetings

(a) All of the above (b) 1 and 2 only
(c) 1 and 3 only (d) None of these

Q.26 Maximum amount of budget is expended during which process
(a) Initiation (b) Planning
(c) Execution (d) Closing

Q.27 Which of the process act as an umbrella process for all other processes?
(a) Initiation process
(b) Planning process
(c) Execution process
(d) Monitoring & control process

Q.28 The project execution phase is done with the aim of
1. Making the necessary change in the product.
2. To make action into operation
3. To change project charter
4. To achieve deliverables

(a) All of the above (b) 1, 2 and 3
(c) 1, 2 and 4 (d) 2 and 4 only

Q.29 Which of the following are correct?
1. Project execution is the most difficult part of a project where real risk are faced off.
2. Stakeholders are responsible for project execution

(a) Both 1 and 2 (b) 1 only
(c) 2 only (d) None of these

Q.30 In regard of project management various monitoring and control tools are:
1. WBS 2. PERT/CPM
3. Line of balance 4. Milestone chart

(a) All of the above (b) 1, 2 and 3
(c) 2, 3 and 4 (d) 3 and 4 only

Q.31 In a project various schedule compression techniques include:
1. Line of balance 2. Crashing
3. Fast tracking 4. Bar chart

(a) All of the above (b) 1, 2 and 3
(c) 2 and 3 only (d) 2 and 4 only

Q.32 Cost variance in a project is the difference between
(a) Earned value & planned value
(b) Earned value & actual cost
(c) Planned value & actual cost
(d) Cost variance & cost performance index

Q.33 Which among the following is not a cost estimating tool for existing (old) product :
1. Parametric 2. Analogous
3. Bottom up 4. Monte carlo

(a) All of the above (b) 1 and 2
(c) 3 and 4 only (d) 1 only

Q.34 In cost estimate of a project, the order of cost estimate (i.e. rough G ± 50% or definite G ± 10%) during successing stages of process
(a) Increases (b) Decreases
(c) Same always (d) Can't be estimated

Q.35 In a precedence diagram the arrow normally indicates
(a) Finish of one to the start of another activity
(b) Start to start of other activity
(c) Finish to finish of other activity
(d) Finish to start of same activity

Q.36 Which of the following are correct?
(i) Resource smoothing is done after resource leveling
(ii) Resource leveling may change project duration

(a) Both (i) & (ii) (b) None of these
(c) (i) only (d) (ii) only

Q.37 Resource levelling is done when
1. To optimise the resource utilisation
2. To generate the uniform demand of resources

Which of these statements is/are correct?
(a) Both 1 and 2 (b) 1 only
(c) 2 only (d) None of these

Q.38 Your cost performance index tells the monetary value achieved for every rupee spent. It value
(a) always < 0 (b) always > 0
(c) 0 < value < 1 (d) can be <, =, > 0

Q.39 Which of the following are correct regarding crashing?
(i) It reduces the project duration
(ii) It increases the cost
(iii) It increases the project duration
(iv) It reduces the cost.

(a) (i) and (ii) (b) (ii) and (iii)
(c) (i) and (iv) (d) (iv) only

Q.40 In a project poor execution may be the effect of
1. Poor leadership
2. No conflicts management system
3. Demotivated workers
4. Multitasking

(a) All of the above (b) 1, 2 and 3
(c) 1, 3 and 4 (d) 2 and 4 only

Q.41 The primary goal of executing process is
(a) Plan the procurement
(b) Direct and manage project execution
(c) Monitoring the various process
(d) communicate details to sponsor

Q.42 Scheduling helps in project execution phase by
(a) By showing sequence & interdependencies of activities with one another.
(b) By estimating the critical path of the method
(c) By creating a WBS
(d) It don't help

Q.43 A positive cost variance means project is
(a) Oven budget (b) Under budget
(c) Equal to budget (d) Can't be predicted

Q.44 A positive schedule variance indicate project is
(a) a head of schedule
(b) being the schedule

(c) On schedule
(d) Can't be predicted

Directions : Each of the next items consists of two statements, one labelled as 'Assertion (A)' and the other as 'Reason (R)'. Examine these two statements carefully and select the answers to these items using the codes given below:

Codes:
(a) Both Assertion (A) and Reason (R) are individually true and Reason (R) is the correct explanation of Assertion (A).
(b) Both Assertion (A) and Reason (R) are individually true but Reason (R) is NOT the correct explanation of Assertion (A).
(c) Assertion (A) is true but Reason (R) is false.
(d) Assertion (A) is false but Reason (R) is true.

Q.45 **Assertion (A):** Crashing is done only for critical activities.
Reason (R): Critical activities duration if reduced reduces the total project duration. So more resources are allotted to these critical activities.

Q.46 **Assertion (A):** Most conflict in a project occurs during the execution phase.
Reason (R): Execution takes most of the resources & is longest time consuming process.

Q.47 Windsor LTD is considering a project, which will involve the following cash inflows and outflows:

	₹. '000
Initial Outlay	**(400)**
After 1 year	40
After 2 year	300
After 3 year	300

What will be the NPV (net present value) of this project if a discount rate of 15% is used?
(a) +₹ 60.8 k (b) –₹ 60.8 k
(c) +₹ 240 k (d) +₹ 460.8k

Q.48 The reason that the schedule performance index (SPI) is shown as a ratio is to
(a) Enable a detailed analysis of the schedule regardless of the value of the schedule variance
(b) Distinguish between critical path and noncritical path work packages
(c) Provide the ability to show performance for a specified time period for trend analysis
(d) Measure the actual time to complete the project

Q.49 The key to effective cost control is
(a) Using earned value to forecast project status
(b) Focusing on projected expenditures and actively networking with key stakeholders to ensure funds will be available as requested
(c) Informing stakeholders of the project's cost status
(d) Managing the approved cost baseline and any changes to it.

Q.50 Project execution must be compared, and deviations must be measured for management control according to the
(a) Scope baseline
(b) Performance measurement baseline
(c) Schedule baseline
(d) Control system

Q.51 A project with both Schedule Performance Index (SPI) and Cost Performance Index (CPI) of 0.80. The project is currently:
(a) Ahead of schedule and under budget
(b) Behind schedule and under budget
(c) Ahead of schedule and over budget
(d) Behind schedule and over budget

Q.52 A project with Earned Value (EV) = \$ 250, Actual Cost (AC) = \$200 and Planned Value (PV) = \$350. What is the Schedule Performance Index (SPI)?
(a) 1.25 (b) 0.80
(c) 0.71 (d) 1.40

Q.53 With reference to the diagram below, it can be inferred that the project is currently :

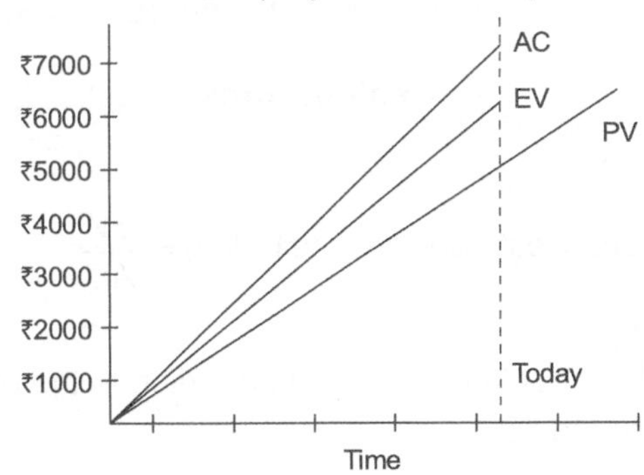

(a) Ahead of schedule and under budget

(b) Ahead of schedule and over budget
(c) Behind schedule and under budget
(d) Behind schedule and over budget

Q.54 Crashing of a project is the process of reducing the :
(a) Cost of the proejct
(b) Time of the project
(c) Material of the project
(d) Labor of the project

Q.55 Choose the correct statement?
(a) Indirect costs decrease as project duration increases and vice versa
(b) Indirect costs increase as project duration increases and vice versa
(c) Indirect costs remains the same as project duration increases
(d) No relationship between Indirect cost and duration of the project

Answers

1. (c)	2. (a)	3. (b)	4. (d)	5. (c)
6. (d)	7. (b)	8. (b)	9. (b)	10. (b)
11. (d)	12. (a)	13. (a)	14. (a)	15. (a)
16. (a)	17. (b)	18. (d)	19. (a)	20. (a)
21. (d)	22. (c)	23. (d)	24. (b)	25. (d)
26. (c)	27. (d)	28. (c)	29. (b)	30. (a)
31. (c)	32. (b)	33. (c)	34. (b)	35. (a)
36. (c)	37. (a)	38. (b)	39. (a)	40. (a)
41. (b)	42. (a)	43. (b)	44. (a)	45. (a)
46. (a)	47. (a)	48. (c)	49. (d)	50. (b)
51. (d)	52. (c)	53. (b)	54. (b)	55. (b)

Explanations

2. (a)

CPI (Cost Performance Index) = $\frac{EV}{AC}$

3. (b)

Fast tracking is a way to accelerate the project schedule.

17. (b)

Case of construction project

= Direct + Indirect cost

Direct cost is a function of time and indirect cost may or may not be a function of time.

Cost – Slope = $\frac{\Delta C}{\Delta t}$ = function of time

= change in cost per unit change in time

≡ extra cost which will incur

18. (d)

Project team salaries are classified as direct costs.

20. (a)

Often project cost estimates are prepared after budgetary approval is provided. However, activity cost estimates should be prepared before the budget is complete.

21. (d)

The control costs process involves monitoring the project's status to update the project costs and managing changes to the cost baseline. Its benefit is that it provides the means to recognize variance in order to take corrective action and minimize risks. Therefore, effective management of the approved cost baseline and any changes is imperative.

22. (c)

Controlling and monitoring is done to complete the project by the stipulated data.

24. (b)

Project management plan is outcome of planning process.

29. (b)

Not all stakeholders are responsible for project execution :

Ex. : Sponsor don't execute the project team members do.

32. (b)

Cost variance = Earned value – actual cost.

34. (b)

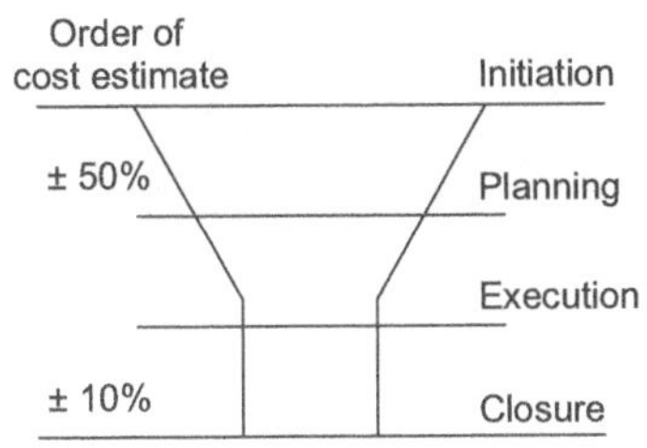

36. (c)
Resource smoothing thing is done after resource leveling. Smoothing don't change duration of project.

38. (b)

$$\text{CPI} = \frac{\text{Earned value}}{\text{Actual cost}}$$

$\because$ Earned value & actual cost both are positive

$\therefore$ CPI > 0

39. (a)
Crashing is done to reduce project duration by increasing the resources for critical activities.

41. (b)
The main aim of executing process is to direct and manage project execution for completion of deliverables

42. (a)
Interconnection is very necessary to define which activity follows the other. Scheduling also reduces time by sequencing not connected activities in parallel.

43. (b)

CV = EV − AC

= + → under budget

− → Over budget

44. (a)

SV = EV − PV

= + → ahead of schedule

− → behind the schedule

47. (a)

$$\text{NPV} = (40 \times 0.87) + (300 \times 0.76) + (300 \times 0.66) - 400$$

$$= +₹\,60.8\text{k}$$

48. (c)
Provide the ability to show performance for a specified time period for trend analysis

Because schedule performance index (SPI) and cost performance index (CPI) are expressed as ratios, they can be used to show performance for a specific time period or trends over a long-time horizon.

49. (d)
The control costs process involves monitoring the project's status to update the project costs and managing changes to the cost baseline. Its benefit is that is provides the means to recognize variance in order to take corrective action and minimize risks. Therefore, effective management of the approved cost baseline and any changes is imperative.

50. (b)
The PMB is an approved, integrated scope-schedule-cost plan for the project work against which project execution is compared in order to measure and manage performance. It includes contingency reserve but not management reserve. It typically integrates scope, schedule, and cost parameters of the project, but it may also include technical and quality parameters.

51. (d)

CPI < 1 = over budget

and SPI < 1 = behind schedule

So the project is both "behind schedule and over budget".

52. (c)

$$\text{SPI} = \frac{EV}{PV} = \frac{250}{350} = 0.71$$

$$\text{SV} = \$1000 - \$800 = \$200$$

Note that the Actual Cost (AC) is not used in the calculation.

53. (b)

$$\text{SPI} = \frac{EV}{PV} > 1$$

$$\text{CPI} = \frac{EV}{AC} < 1$$

■■■■

7 PROJECT CLOSURE & TERMINATION: Post Audit

7.1 PROJECT EVALUATION

- To evaluate means to assess. Thus, project evaluation is an assessment of the project during the course of implementation. It is an interim assessment at major milestones. Evaluating the project at major milestones during the course of implementation will steer the project in the right direction. Project evaluation is an attempt to determine if the overall status and progress of the project is acceptable as compared to what was planned earlier and if the objectives are being achieved. The ultimate aim of project evaluation is to bring about all-round improvements in project planning and execution.

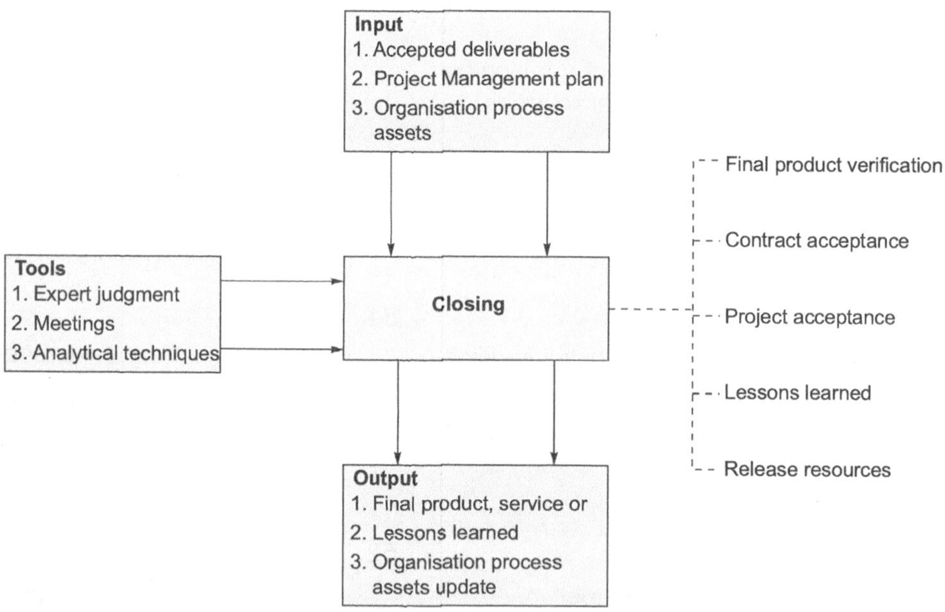

7.2 EVALUATION METHODS

- Smaller projects may be evaluated by direct observation and by getting first-hand information from the field staff, specialists and others connected with the project implementation. Evaluation based on periodic written reports is done for major projects, the written reports may be supplemented with charts, graphs and illustrations to add value to the reports.

Post Project Evaluation (Post Audit)

- Post project evaluation is also known as Post Audit or Post Completion Audit. It is an evaluation of the project after its completion. At the time of project appraisal, estimates are made for project cost and for project completion time. After the project is completed, project audit is carried out to assess the actual project cost and the actual time taken for project completion. Thus, while project appraisal is an estimate for the 'future', post project evaluation is an assessment of the 'past'. Ideally, a project will be considered successful if it gets completed on time, within the budget and gives the expected level of performance. In respect of public projects, apart from studying the variations in project cost and project completion time, post project evaluation also makes an assessment of the actual social cost-benefit factors and the extent up to which the project's objectives are achieved.

Objectives of Post Audit

- Experience is the greatest teacher. The knowledge gained during the process of project implementation is of great value to all concerned. The project owner and the project management team learn a lot during the course of project implementation and they become a store house of knowledge which they can share with others or use them in their future ventures.
- By carrying out post project evaluation, the financial institution that has funded a project can identify the pitfalls, if any, in its project appraisal and the control mechanism that should have been followed for effective project monitoring.

The objectives of post audit can be summed up as under:

- Building up an information base help proper estimation of project cost and time.
- Educating all those concerned with the project about the realities of project management.
- Establishing correct time-cost relationship.
- Creation of appropriate standards for work based on suitable work techniques.
- Sharing of project audit information among all concerned, in order to build up better understanding and better comprehension of the project and its problem areas so that lapses could be avoided in future.

Phases of Post Audit

- Post audit is carried out in two phases, *viz.*, immediately after the completion of the project and after the lapse of sometime (say 2 or 3 years) since the completion of the project.

  Post audit that is carried out immediately after project completion is done.

  1. for studying the differences between the actual project cost and the estimated project cost.
  2. for studying the difference between the actual time taken for project implementation and the estimated time.
  3. for locating the areas that have contributed to the variances in project cost and time.
  4. for identifying the reasons for such variances, classifying them into avoidable and unavoidable variances.
  5. for analyzing the steps that could have been taken to avoid the avoidable variances
  6. for analyzing in depth the factors that have caused the unavoidable variances and examining the possibilities of their removal in future by adopting suitable methodology.

  Post audit that is carried out after the lapse of 2 to 3 years is done with a wider perspective. This study is done:

  1. for studying whether the product goals and objectives are achieved.
  2. for knowing whether the project produces products of acceptable quality.
  3. for knowing whether the estimated output is achieved.
  4. for knowing whether the product is accepted by the market and whether the production volume is commensurate with the market share planned to achieve.

The post project evaluation, whether done immediately after the project completion or after the lapse of sometime has the main aim of studying the pitfalls and gray areas and identifying the required corrective measures.

Types of Post Audit

- **Technical evaluation (or Technical Audit):** Technical evaluation refers to the evaluation of quality and quantity of production, the operating costs in production etc. A comparison of these factors is done between what is presented in the feasibility report/detailed project report and what has been the actual achievement. As regards the rated output, certain projects are expected to give the estimated rated output immediately on commissioning. Such projects can be evaluated, for the variations in output if any, immediately after commissioning. In respect of certain other projects, the output will stabilize only after some time. These are the cases where the output depends upon the Based on the points discussed above the objectives of post audit can be summed up as under the quality of raw materials, the skill of the personnel operating the plant etc. The operation of the plant is to be studied by varying these parameters and observing the output till the production is stabilized at the maximum level. Evaluation of output can be done only after this *i.e.,* only after all the inputs are synchronized and the production stabilized.
- Apart from evaluation the actual output *vis-a-vis* the rated output, the other areas of interest are the evaluation of utilities consumed by the plant, like power, fuel, water, steam, consumable, spares etc. An evaluation of these aspects is made and compared with what was estimated/projected at the time of project appraisal. Technical evaluation also includes evaluation of the quality of the output and checking-up if the quality standards as envisaged are reached.
- **Financial evaluation (or Financial Audit):** Financial projections/estimates are made at the time of project appraisal as to the project cost under various heads, operating costs of the project. Maintenance costs, profitability estimates, cash-flow and fund- flow estimates, sources and application of funds for the project etc. Financial evaluation is done to verify whether the actual project cost, operating costs, profitability, cash/fund flows etc., are as per the estimates and projections made at the time of appraisal. While the assessment of actual project cost can be done immediately after the completion of the project, other financial parameters can be studied only after the lapse of say, two to three years since the record of actual operating costs, profits etc. would be available only after the project is in operation for some time.
- **Economic evaluation (or Economic Audit):** Economic evaluation is the most difficult to make since it involves many subjective aspects which are difficult to be quantified. The evaluator should have an 'eye" for identifying the social costs and benefits of the project. Economic evaluation is more relevant for public sector and community development projects since such projects are undertaken with social objective in mind, apart from financial and other objectives.
- A public sector project would have been implemented with the aim of achieving certain social benefits. The social costs and social benefits, though subjective in nature, are quantified using some techniques while doing Social Cost Benefit Analysis (SCBA) at the appraisal stage of the project. The project audit from Social Cost Benefit angle should analyse whether the social benefits envisaged are achieved and whether there are any adverse effects that the project has brought in. Which were not originally foreseen. Such unforeseen adverse effects have the nature of adding to the social cost of the project. Though SCBA at the appraisal stage has grown into a well laid out system of appraisal in the recent past, post project evaluation from social cost-benefit angle is yet in an infant stage.

7.3 PROJECT CLOSURE

One might have delivered the deliverables, but this does not mean your project is complete. Ignoring this process results in incomplete project management at the project manager's end. As closing a project is as important as initiating it,

According to A Guide To The Project Management Body of Knowledge (PMBOK® Guide) "The Project Closing Process Group consists of those processes performed to conclude all activities across all Project Management Process Groups to formally complete the project, phase, or contractual obligations. This process group, when completed, verifies that the defined processes are completed within all of the Process Groups to close the project of phase, as appropriate, and formally establishes that the project or project phase is complete"

Closing Process Group Activities

- Confirm work is done as per requirements.
- Complete proceurement closure
- Gain formal acceptance
- Complete final performance reporting
- Index and archieve records
- Update lesons learned

After delivering the deliverables and receiving formal sign-off from the customer, the project manager must hand over the resources to their respective departments so that they can be sourced for other projects.

Thereafter the following need to be done in the Project Closure.

Procurement or Other Contract Closure: If a third-party vendor or any subcontractors have been working on the project, those contracts need to be closed. Once they have produced deliverables, and you have delivered them to your end customer, the contracts should be closed.

Indexing of the Project Files: Once the project is over, the deliverables have been handed over to the customer, and the customer signs off, the project files should be compiled and conveyed to your entire list of key stakeholders. The archived files could be used in the future.

Lessons Learned Documentation: Once formal sign-off has been received from the customer, the lessons learned from this project should be documented which can be used as a reference for future projects.

Circumstances for change requests that were accepted or rejected, history of the schedule control, and cost control could be assessed with the lessons learned. It serves as an important document in case of project closure.

Project Closure and Termination: Projects are temporary and therefore must come to an end. When planning a project we schedule for project completion and plan closure accordingly. However, a project may close for variety of reasons.

Gray and Larson (2008) identify five circumstances for project closure:

1. **Normal closure:** The common condition of project closure is when the project is completed as planned. This is when the project objectives are achieved, the client accepts the project and normal project closure commences.
2. **Premature closure:** Many projects do not achieve all their deliverables or are not given the opportunity to so. Instead, they are closed prematurely by eliminating elements of the project originally identified in the project scope.

   This could be for the reason of costs, where the client reduces funds on the project or the project has already consumed the budget. Premature closure also occurs when the project is of strategic importance and must be delivered earlier than expected, such as a new product launch. Delaying the product until the original completion date may result in opportunity lost for the client.

3. **Perpetual projects:** Conversely, some projects never seem to end. These are projects that have had numerous delays, setbacks and problems. Perpetual projects also suffer from endless scope creep, add-ons and changes.The problem with these types of projects is that they never achieve their goals or objectives, due to the changes and consistent scope creep. This becomes highly frustrating for the project manager and the project team. It will also be highly frustrating for the client as they do not see the objectives of the project being achieved, despite the consistent request for changes. At some point the project manager needs to fix the scope and plan for closure. Redefining the project scope so project closure is forced, limiting budget or resources, or setting a time limit can do this. As a result, any additions that the client requests can be viewed as a second phase of the project, rather than the project being perpetual
4. **Failed projects:** Far too often projects close because they have failed. There are a number of causes for project failure. It is not uncommon for the client to run out of funds, thus permanently killing the project.
5 **Changed Priority :** The organization's strategy can change during the course of time according to the market environment changes. The change in such priorities too result in project closure.

Mereditch and Mentel's classification of Project Termination: A project can be said to be terminated when work on the substance of the project has ceased or slowed to the point that further progress is no longer possible

There are four fundamentally different ways to close out a project: extinction, addition, integration, and starvation.

Termination by Extinction The project may end because it has been successful and achieved its goals.

The project may also be stopped because it is unsuccessful or has been superseded.

A special case of termination by extinction is "termination by murder" which can range from political assassination to accidental projecticide. When a decision is made to terminate a project by extinction, the most noticeable event is that all activity on the substance of the project ceases

Termination by Addition: If a project is a major success, it may be terminated by institutionalizing it as a formal part of the parent organization.Project personnel, property, and equipment are often simply transferred from the dying project to the newly born division.The transition from project to division demands a superior level of political sensitivity for successful accomplishment.

Termination by Integration: This method of terminating projects is the most common way of dealing with successful projects, and the most complex.The property, equipment, material, personnel, and functions of the project are distributed among the existing elements of the parent organization

In general, the problems of integration are inversely related to the level of experience that the parent or client has had with the technology being integrated and the successful integration of other projects, regardless of technology

Termination by Starvation: This type of project termination is a "slow starvation by budget decrement".

There are many reasons why senior management does not wish to terminate an unsuccessful or obsolete project such as it may be politically dangerous to admit that one has championed a failure or terminating a project that has not accomplished its goals is an admission of failure

Objective Brain Teasers

Q.1 The project manager is making sure that the product of the project has been completed according to the project management plan. What part of the project management process is he in?
(a) Planning
(b) Executing
(c) Monitoring and controlling
(d) Closing

Q.2 All projects, are officially initiated with a _____ and officially completed with a _____.
(a) Project charter, certificate of acceptance
(b) Detailed project report and certificate of acceptance
(c) Project charter and closure report
(d) Detailed project report and closure report

Q.3 The different steps in verification of scope of work are
1. Approved design 2. Material acceptance
3. Equipment 4. Quality control
(a) 1, 2 and 3 (b) 1, 3 and 4
(c) 2, 3 and 4 (d) All of these

Q.4 Project handover process consist of
1. Training
2. As-built drawings
3. Operating Manual
4. Transfer of responsibility
5. Warranties
(a) 1, 2, 3 and 5 (b) 2, 3, 4 and 5
(c) 1, 3 and 5 (d) 1, 2, 3, 4 and 5

Q.5 The characteristics and feature of the project commissioning and handover process are
1. Receive the deliverables and documents
2. Verify the scope of work
3. Test and commission of project
4. Handover project to client
(a) 1, 2 and 4 (b) 2, 3 and 4
(c) 1, 3 and 4 (d) All of these

Q.6 _____ as, the advancement of an installation from the stage of static completion to full working order and achievement of the specified operational requirement.
(a) Project execution
(b) Commissioning
(c) Certificate of acceptance
(d) Project closure

Q.7 A schedule performance index *SPI* of 0.76 means:
(a) you are over budget.
(b) you are ahead of schedule.
(c) you are only progressing at 76 percent of the rate originally planned.
(d) you are only progressing at 24 percent of the rate originally planned.

Q.8 Bottom up estimation is done under the following conditions on a project:
(a) When there are a lots of previous projects that can be used as a reference to prepare a budget
(b) When the project manager's previous experience can be utilized to make a budget
(c) When we do not have enough time to prepare a project budget
(d) When we do not have previous project information to prepare a budget

Q.9 Why proper closing of the project is necessary?
1. To release the resources
2. To let everyone of stakeholder know about the project closure
3. To stop further expenses
(a) All of the above (b) 1 and 2 only
(c) 2 and 3 only (d) 1 and 3 only

Q.10 You are a stakeholder (financer) of a project. Your project can be closed by following situations
1. You are out of money
2. Your project manager leaves
3. Your project objective is complete
4. Your are wanting to close the project.
(a) All of the above (b) 1, 2 and 3
(c) 1, 2 and 4 (d) 1, 3 and 4

Q.11 Project is officially closed when
(a) Deliverables are manufactured
(b) Deliverables are accepted
(c) Project manager leaves project
(d) Lessons learned has been performed

Q.12 What are different input to a project closure ?
1. Project management plan
2. Accepted deliverables
3. Organisation process assets

(a) All of the above (b) 1 and 2 only
(c) 2 only (d) 1 and 3 only

Q.13 Project is closed when
(a) administrative closure is complete
(b) contact closure is complete
(c) deliverable are accepted
(d) lessons learned are performed

Q.14 Lessons learned at the end of each project is done so that
1. Next people can learn from the project
2. To know what all can be made better
3. To criticise the non-functional members
(a) All of the above (b) 1 and 2 only
(c) 2 and 3 only (d) 2 only

Q.15 Audit of a project can be done
1. Before completion of project
2. After completion of project
3. Before project starts
(a) All of the above (b) 1 and 2
(c) 2 only (d) 1 only

Q.16 What are objectives of post audit of project?
1. Enlighten all concerned people
2. To check product is operating as promised
3. Creation of standards of work.
(a) All of the above (b) 1 and 2
(c) 2 and 3 (d) None of the above

Q.17 In order to administrative closure it is necessary that
1. Deliverables are accepted
2. Deliverables are not accepted but are as per the scope, project charter & statement of work.
3. Contract closure is done
(a) All of the above (b) 1 and 2
(c) 1 and 3 (d) 1 Alone

Q.18 Scheduling an audit play an important role as
1. It fixes the time for audit
2. Delaying in project audit may lead to loss of money and opportunity cost.

Which of these statements is/are correct?
(a) Both 1 and 2 (b) 1 Alone
(c) 2 Alone (d) None of these

Q.19 Who is responsible for updating, managing & developing lesson learned documents?
(a) Stakeholder (b) Project manager
(c) Project team (d) customer

Q.20 A project ends when
1. Objective achieved
2. Project manager closes the project
3. Client closes the project.
4. Objective of project can't be achieved.
(a) 1, 2, 3 and 4 (b) 1 and 3
(c) 1, 2 and 4 (d) 1, 3 and 4

Q.21 Whose responsibility is to balance the constraints in a project?
(a) Project sponsor (b) Project manager
(c) Project team (d) Technical team

Q.22 In a project to create a new mobile phone your sponsor asks you that the launch date is brought forward. What will be its impact?
(a) cost increases
(b) cost decreases
(c) scope increases
(d) cost remains unaffected

Q.23 All the question of a contractor are assured
(a) Kickoff meeting (b) bidder conference
(c) project charter (d) issue log

Q.24 When a project is not able to sustain due to lack of budget and it terminates, this type of termination is called?
(a) Termination by addition
(b) Termination by integration
(c) Termination by starvation
(d) Termination by extinction

Q.25 You are a project manager, you have completed the project as per the baseline but the customer is not satisfied, He wants more to be done without pay, what should you do?
(a) Quit the project
(b) Close the project and perform lessons learned.
(c) File a case against the customer
(d) Take more time and do as customer is satisfied.

Q.26 When a project is such that its scope is changing many times and it becomes never ending? Then this type of project closure is called
(a) Normal closure (b) Premature closure
(c) Perpetual closure (d) Failed project

Q.27 Which among the following can be done after project closure?

(i) Pre auditing (ii) Post auditing
(iii) Lessons learned (iv) Contract closure
(a) All of the above (b) (i) and (iii) only
(c) (ii) only (d) (ii) and (iii) only

Q.28 You as a project manager are in execution phase of project in weak matrix organization. At some point you get to know that additional human resources are required, whom would you request
(a) sponsor
(b) You yourself have authority
(c) Functional manager
(d) Project stakeholder

Q.29 Which among are generally elements in project charter?
(i) Formal authorization of "Go" decision for project
(ii) Product description
(iii) Business need of the project
(iv) Project manager identification.
(a) All of the above (b) (i), (ii) and (iv)
(c) (i), (iii) and (iv) (d) (i), (ii) and (iii)

Q.30 What does RAM stands for in project management?
(a) Random Accounts Management
(b) Responsibility Assignment Matrix
(c) Risk Awareness Management
(d) Repeatable Applied Milestones

Q.31 What does the S-curve describes in a project management?
(i) Growth of earned value
(ii) Cumulative work that is to be done
(iii) A comparison in earned value and planned value for earned value analysis
(iv) Time remaining to complete the project
(a) All of the above (b) (i), (ii) and (iii)
(c) (i) & (iii) (d) (ii), (iii) & (iv)

Q.32 Resource Planning includes which of the following:
(i) Human resources
(ii) Material resources
(iii) Capital equipment
(a) All of the above (b) (i) and (ii)
(c) (ii) only (d) (i) and (iii)

Q.33 Scheduling of activities in a project management requires which of the followings
(i) Sequence of activities
(ii) Duration of activities and project
(iii) Resource estimation for all activities completion
(iv) Risk analysis of each activity.
(a) All of the above (b) (i), (ii) and (iii)
(c) (i), (ii) and (iv) (d) (i) and (iv) only

Directions : Each of the next items consists of two statements, one labelled as 'Assertion (A)' and the other as 'Reason (R)'. Examine these two statements carefully and select the answers to these items using the codes given below :

Codes:
(a) Both Assertion (A) and Reason (R) are individually true and Reason (R) is the correct explanation of Assertion (A).
(b) Both Assertion (A) and Reason (R) are individually true but Reason (R) is NOT the correct explanation of Assertion (A).
(c) Assertion (A) is true but Reason (R) is false.
(d) Assertion (A) is false but Reason (R) is true.

Q.34 **Assertion (A):** It is a good practice to do contract closure first than project closure
Reason (R): In this manner vendors get time to submit the final bill before closure of project.

Q.35 **Assertion (A):** Product or service created by a project can be non-temporary in nature.
Reason (R): Every project is temporary in nature.

Q.36 Today, the primary cause of why projects are not completed on time and within cost and are terminated early is due to
(a) Contractual issues
(b) Complexity of the project
(c) An increase in the allocated time
(d) Behavioral-oriented reasons

Q.37 When managing projects, one important technique is capturing lessons learned from previous projects to improve the organization's project management process. Therefore, in planning the project, it is important to plan the closing procedures, particularly to review the
(a) Risk responses that were used
(b) Checklists for risk identification
(c) WBS and scope documents
(d) Team member's performance feedback

Answers

1. (d)	2. (c)	3. (d)	4. (d)	5. (d)
6. (b)	7. (c)	8. (d)	9. (a)	10. (d)
11. (b)	12. (a)	13. (a)	14. (b)	15. (b)
16. (a)	17. (a)	18. (a)	19. (a)	20. (d)
21. (b)	22. (a)	23. (b)	24. (c)	25. (b)
26. (c)	27. (c)	28. (c)	29. (a)	30. (b)
31. (c)	32. (a)	33. (b)	34. (a)	35. (b)
36 (d)	37. (b)			

Explanations

10. (d)
Project is not closed if any of team member or project manager leaves the project.

11. (b)
Project can be closed after deliverables are accepted.

13. (a)
A project closes officially after administrative closure.

21. (b)
A project manager is responsible for integration of different areas and to take the project to success.

22. (a)
If launch date is brought near, its time decreases and mare resources are needed to complete task in a given time.

23. (b)
Bidder conference is meeting prior to contract bidding where all the bidders can ask there questions openly. This is done so as to avoid any favouritism to any bidder.

24. (c)
Termination by starvation is when a project is not able to sustain due to lack of budget.

25. (b)
As a project manager your work is to finish the project as per the baseline & scope agreed. Once the project is completed as per baseline you can close the project, though sometimes customer might not be satisfied.

26. (c)
If a project becomes never ending due to frequent change of scope attempts should be made to close it.

27. (c)
Pre auditing, lessons learned and contract closure has to be done before projects administrative closure.

28. (c)
In a weak matrix function manager are responsible for different resources.

31. (c)
S-Curve generally depicts the growth of earned value as project progress. It is also used in earned value analysis.

33. (b)
Risk analysis is not done in scheduling.

36 (d)
Behavioral reasons, rather than quantitative reasons, account for more project terminations because it is much more difficult to manage people than things. Issues such as poor morale, poor human relations, poor labor productivity, and no commitment from those involved in the project combine to thwart project success in many industries.

37. (b)
Checklists are a tool and technique of the identify risks process and include risks encountered on similar, previous projects identified through the lessons learned process. The project team will review the checklist as part of the identify risks process as well as during closeout. To help others in the future, the team will add to the list as necessary based on its experience.

8 Tender and Contract

8.1 PROJECT CONTRACT

- A contract is a legal agreement. It is an exchange of promises by two or more persons/organisations.
- Section 2 *(b)* of the Indian Contract Act, 1872 defines contract as "an agreement enforceable by law". The definition given by the Indian Contract Act has two distinct parts, *viz.*,
  (a) a contract is an agreement
  (b) the agreement must be enforceable by law.
- Though contract is necessarily an agreement, but all agreements are not contracts. Only those agreements that are enforceable by law are contracts. For an agreement to become enforceable by law, it must satisfy certain essential elements called "essentials" of valid contract".
- The contractor is a person, a firm or a company who undertakes contract.

Essential features of a valid contract are as under:

- There must be an offer and an acceptance of that offer (Public notice inviting tenders for the execution of a work on contract basis is an offer and if a contractor submits his tender documents in response to the public notice and in accordance with the terms and conditions stipulated, it amounts to acceptance of the offer.)
- There must be an intention to create a legal relationship on the part of the parties to the contract.
- The consent of the parties to the contract shall be free and genuine.
- The parties to the contract shall be competent persons. (For example, minors and lunatics do not have the capacity to enter into any contract.)
- There must be a lawful consideration between the parties to the contract.
- The object of the contract must be lawful and not one which the law disapproves.
- The contract must not have been declared as void by any law in force in the country.
- The meaning of the contract must be certain without any room for ambiguity or different interpretations.
- The terms of the contract shall be capable of performance.
- A contract shall fulfill the necessary legal formalities.

8.2 TYPE OF CONTRACTS

Broadly contracts can be divided into two types. They are:

(a) Turnkey contracts

(b) Non-turnkey contracts

(a) Turnkey Contracts

- In a turnkey contract, the entire responsibility of project execution is entrusted to the contractor. It is as if the owner comes into the picture only when the project is completed and he turns the key of the plant to start production. Till such time the project reaches a 'ready-to-start' stage, the contractor takes care of all aspects of project execution.

- Since the contractor of a turnkey project is expected to do everything right from scratch, the scope of contract covers all areas of the project, *viz.,* design, engineering, construction, structural work, supply and erection of plant and machinery, supply of spares, testing and commissioning. Since the entire work is done by a single contractor, turnkey contract agreements invariably have a clause on performance guarantee.
- Turnkey contracts may be for the whole project or even for a sub-unit of a main project. Turnkey contract for a sub-unit is undertaken when such sub-unit is a separately identifiable component of the total project and whose performance can be assessed independent of other units of the project.

(b) Non-turnkey Contracts

- Non-turnkey contracts are preferred when the projects are small sized, the know-how for the project is available with the promoters and when there is a strong, competent and capable project team available with the organisation.
- In non-turnkey arrangement, the total project is divided into suitable work packages. The packages that need to be given to contractors are identified. A contractor may be entrusted with one or more work packages. However, the responsibility for overall supervision and performance of the project will rest with the in-house project department.

The following are some of the non-turnkey contracts:

1. **Lumpsum contract (Fixed price):** This is a single fixed price contact. In this contact, contractor agrees to perform specified job for fixed sum. The owner provides the contractor exact specification of the work. In this contract following are the advantages of the fixed price contract.
   (a) Owner in aware of the cost of the project before the project construction starts.
   (b) It avoids a lot of details and accounting by both owner and contractor
   (c) Contractor gets free hand to execute the work
   (d) In this contract is used with design contract method of delivery, contractor gets opportunity to use value engineering.

   In this contract following are the disadvantages of the fixed price contract.
   (a) It is very difficult to accommodate any change and specification.
   (b) This contract is as good as the accuracy of the contract document. If errors exist in the contract document, the contract need to be renegotiated and hence more risk in involved from the owner side.
   (c) In the case of unforeseen hazard during the construction, contractor may be put in adverse situation.

   Payment by the owner can be carried out in lump-sum contract as total amount at time or percentage of total cost after finishing certain amount of work. For example suppose a water tank was awarded as lump-sum contract. It can be said that 15% of amount will be paid after construction of foundation 50% of total amount will be paid after construction of staging and 80% of the total amount will be after the construction of the tank. Rest will be paid only when the tank becomes operational. Lump-sum contract in often used in sub-contracting for labour contract. In our country many labourers for excavation, plastering work with this method.
2. **Unit price contract:** In this type of contract, the price is paid per unit of the work carried out. There are different variations of this type of contract. Some of them are mentioned below.

Bill of quantities contract: In this type of contract owner provide the drawing, quantities of work to be done and specification. The contractor bid based on the unit cost of the items of construction. The contractor overhead, profit and other expanses can be included in the unit cost of the item of work. Sometimes contractor quotes the unit price of the work and lump-sum amount separately as profit overhead. The estimated quantities of the work to be done called Bill of the quantities is fixed.

This type of construction is usually followed in government sector for large infrastructure construction. This type of contract provides owner a competitive bid. Disadvantages of the methods are:

(a) Owner needs to measure the quantity of work done in the field, hence requires owner presence at the site.

(b) Final price of the construction is not known precisely until last price of work is completed. If there is significant difference between the estimated quantities and the reality of the situation, owner is put in adverse situation. Mistaken quantities is called unbalanced bid. Significant unbalanced bid now considered as unethical.

Schedule of rate contract: Many a time, the quantity of work to be executed is not known before. Contract is signed based on the unit cost of the item of work. Generally more items are inserted in the contract than to be executed because it becomes sometimes difficult to exactly specify all the items. There is no guarantee that all the items mentioned will be used in the construction. This type of contact are widely used in underground work, flood control and road constructions. Advantages and disadvantages of this type of contract in the same as the bill qualities contract. There are other variations of schedule of rate contract where unit price plus profit is charged as the cost.

(c) **Cost plus contract:** In this contract, the payment is made based on the work carried out plus the free which includes overhead, profit etc. Sometimes a cap is put on the type of contract by provided maximum and minimum cost limit such as guaranteed maximum cost contract. If project cost exceed this limit, contractor is responsible for that.

Sometimes incentive clause is also included if the contractor bring the project before certain specified limit.

The advantages of this type of contract is that considerable overlap is provide between design and construction. Hence the project can be executed in the fast-tract basis. This contract is suitable for the work where it is difficult to define the task to be done before the awarding the contract.

3. **Item rate contract:** Item rate contract is also known as unit price contract or schedule contract. A contractor undertakes the execution of work on an item rate basis. He is required to quote rate for individual item of work on the basis of schedule of quantities (i.e., bill of quantities) furnished by the department. The amount to be received by the contractor, depends upon the quantities of work actually performed. The payment to the contractor is made on the basis of the detailed measurements of different items of work actually executed by him.

Suitability:

The item rate contract is most commonly used for all types of engineering works of the government undertakings including railway department. It is suitable for works which can be distinctly split into various items and quantities under each item can be estimated accurately.

8.3 PROJECT TENDER

Tender is an offer in writing by the tenderer (the person who offers the tender) to execute some specified work or to supply some specified good at certain rates within a fixed time frame under certain conditions of agreement.

Tender Notice or Notice Inviting Tender (NIT)/ RFP (Request For Proposal)

The tender notice is a public notice and is published in leading newspapers and e-portals. This is a popular method used by Government Departments and Public Sector Enterprises. The tender notice contains the following information:

(a) Name of the organisation calling for tenders and the designation of the officer concerned who is inviting tenders.

(b) The nature of work, the place of work and the estimated cost of work.

(c) Cost of tender documents and drawings/plans.

(d) Last date and time of receipt of tenders, place and availability of tender forms.

(e) Type of specifications and time of completion.

(f) Amount of Earnest Money Deposit (EMD) to be paid and the method of payment of EMD.

(g) Amount of security deposit to be deposited by the successful tenderer.

(h) Date, time and place of opening tenders.

Tender Documents

Tender documents are meant to keep the tenderers informed about the general and specific conditions applicable for tenderers. Tender documents usually consist of the following:

(a) A letter of invitation to the tenderers

(b) Specimen tender form

(c) General instruction to the tenders

(d) Details of civil/structural work along with complete set of civil/structural drawings.

(e) Details and specifications of machinery/equipment to be supplied, if any

(f) Draft contract agreement

(g) Arbitration authority who will decide in case of dispute

(h) Time schedule for completion of work

(i) Amount of earnest money to be deposited and the form in which it is to be deposited.

Pre-Qualification of Tenderers

- Pre-qualification of tenderers is done only in bigger and complex projects. The objective of prequalification is to ensure that only competent tenderers participate in the tenders. The following information is gathered from the intending tenderers for deciding as to whether they are competent enough to participate in the tenders:
- The financial capacity of the contractors for carrying out the work.
- Infrastructure available with the contractors in terms of machinery, equipment, qualified personnel, man-power etc.
- Previous experience in having handled projects of similar nature. After assessing the qualifications of the intending tenderers, those who are found competent are chosen and only the selected persons are allowed to participate in the tenders.

Earnest Money Deposit (EMD)

- It is the amount to be deposited by all the tenderers when they submit their tender. The EMD amount varies from 1% to 3% of the tender value. Once the contract is finalized, the EMD remitted by all the unsuccessful tenderers are returned back. The EMD of the successful tenderer is retained as a measure of caution so that he will not withdraw his offer. In the event of the successful tenderer withdrawing his offer, or refusing to take up the work, the EMD retained will be forfeited. A tenderer can withdraw his offer before acceptance of the tender. Once a particular tender is selected and communicated and if the successful tenderer withdraws his offer after the communication, he has to forego his EMD.
- The EMD is payable normally in the form of pay-order/Demand Draft from a nationalized/ scheduled bank. When the EMD amount is large, EMD is also accepted in the form of Bank Guarantee from a nationalized/scheduled bank.

Security Deposit

- Security Deposit is the amount to be deposited by the successful tenderer after the contract is finalized. Normally it is about 10-15% of the total value of the contract. This amount is collected as safety measure so that the contractor (successful tenderer) fulfills all the terms and conditions of contract and completes the work in accordance with the terms of the contract. Security deposit amount will be forfeited in case the contractor fails to fulfill any of the terms of contract. Security deposit amount will be refunded to the contractor on completion of the work and after observing the work for defects if any till the 'defects liability period' which is normally around six months.

Acceptance of Tenders

- On the appointed date, the sealed envelopes containing tenders and E.M.Ds. are opened. The envelopes containing E.M.Ds are first opened and if found in order, the other envelopes containing tender documents are opened. Opening of tenders is usually done in the presence of the tenderers.
- The valid bids received are scrutinized and evaluated by a committee of experts designated for purpose. The tender evaluation committee will have representatives from all the departments illegal department, finance department, operations department, purchase department etc.) and special invitees who have specialized knowledge in the concerned field. Before accepting one among the tenders received, the following aspects are to be given consideration:
  (a) The financial capacity of the tenderer
  (b) The technical infrastructure available with the tenderer
  (c) Past performance of the tenderer in executing similar works
  (d) Specifications (both general and technical) quoted by the tenderers and deviations in specifications, if any.

  Apart from the above, the draft contract is to be examined and its terms studied, before accepting a tender.

  The lowest tender is usually accepted provided the tenderer who has quoted the lowest amount! Satisfies the above criteria. For valid reasons, a tender other than the lowest can also be accepted.

Letter of Intent (LOI)

This is a letter issued to the successful tenderer intimating about his/her offer being accepted. LOI is issued before signing of the contract by the two parties. The successful tenderer is required to intimate any pre-condition within the stipulated time. Issue of LOI by the project promoters and acceptance of the same by the successful tenderer does not obviate the need for signing a contract. Since LOI is not a legal document, the two parties have to necessarily execute the contract by signing the contract agreement.

Execution of Deed of Contract

Strictly speaking, a contract, as such, can be oral or in writing. But, if a contract is required by law to comply with certain legal formalities, it must comply with the necessary formalities as to writing, registration and attestation, if necessary. If these legal formalities are not carried out, then the contract is not enforceable at law.

Penalty Clause for Non-Performance of Guarantee Parameters

Apart from the clause on liquidated damages, a penalty clause for non-performance of guarantee parameters is also included in the contract. This clause provides for deduction of a certain predetermined percentage of price from the contract price for non-performance of any guarantee parameters. The percentage of deduction is to be carefully decided in consultation with the consultants and contractors in such a way that it sufficiently covers the loss due to non-performance. The percentage of deduction should be such that it neither affects the clients nor the contractors.

Types of Payments to Contractors

The following are the different modes of payment to contractors:

(a) First and final bill

(b) Running account bill

(a) First and final bill

This method of payment is suitable for very small, petty works. There is only a single payment to the contractor on his completion of the work.

(b) Running account bill (R.A. bill)

Running account bill refers to the account prepared at frequent and regular intervals for payment for the work done or supplies made at intervals. During the progress of the work, the contractor is paid from time to time, while the final payment is made on completion of the work. On a particular date when the bill is prepared, it is prepared for the quantity of work executed till that date and at the rates agreed in the contract. Payments are made for about 90% of the bill amount, retaining about 10%. Thus, there will be many intermediate bills and one final bill.

- Final bill is the last bill raised on completion of the work and in full settlement of the account.
- Sometimes advance payment is made on running account under special circumstances where measurements can not be taken for certain valid reasons, but the work has progressed sufficiently.

DO YOU KNOW?

- **Voucher:** Voucher is a written document as a proof of payment. Normally payment made to the contractor is acknowledged by the contractor by putting his signature on the bill on a revenue stamp. A bill, thus becomes a voucher after payment.
- **Liquidated damages:** One of the important clauses in contract is the schedule for completion of work/schedule for delivery. As we are aware, any delay in implementation over the pre-determined schedule results in time-overrun which may lead to cost-overrun. Moreover, a delayed project means, delay in getting back return on investment which will also affect the financial viability of the project. To ensure that the project implementation is carried out as per the original schedule, a clause known as 'liquidated damages' clause is incorporated in the contracts, which stipulates deduction of certain percentage of payment due to contractor in case of delays. The clause can have provisions to assess the delay on monthly, weekly or even on daily basis for very critical projects and provide for deduction of payment from the R.A. bill for every month/week/day of delay.

8.4 MISCELLANEOUS TOPICS

1. Project Sickness

- The sick industrial companies (Special provisions) Act, 1985, as amended in 1993 defines sick industrial company as an industrial company (being a company registered for not less than five years) which has at the end of any financial year accumulated losses equal to or exceeding its entire network.
- Section 3(1) (ga) of the amendment act, 1993 defines "Net worth' as the sum of the paid-up capital and free reserves, while the term "free reserves' means all reserves credited out of the profits and share premium account, but does not include reserve out of reevaluation of assets, write back of depreciation provisions and reserves created out of amalgamations.
- Government companies having State or Central Government share holdings of 51% or more are kept outside the purview of the Act. Also small scale industrial units and Ancillary units are kept outside the purview of the Act?

Reporting to the BIFR (Board for Industrial & Financial Reconstruction)

- The BIFR as an agency of the Government of India entrusted with the task of determining sickness of industrial companies and assisting in the revival of those that were vival while shutting down the others.
- From Dec 2016, BIFR has since been dissolved and proceedings concerned with sickness and revival been released to National Company Law Tribunal (NCLT) and National Company Law Appellate Tribunal (NCLAT) as per the provision of IBC (Insolvency and Bankruptcy Code). The function and responsibilities of BIFR, therefore, have now been entrusted to NCLT and NCLAT.

Causes of Sickness

- 'Prevention is better than cure' is the proverb that reflects the need for knowing the likely causes of industrial sickness so that one can plan to avoid the same. Just as human beings fall sick by two ways, *viz.* Either born sick or acquiring sickness during growth, an industry can either run into trouble even during the implementation stage itself or develop sickness during its lifetime.
- The causes of sickness can be categorized into two viz., internal causes and external causes. Internal causes are those that are internal to the organisation over which the management of the organisation has control. Sickness due to internal causes can be avoided if the management is shrewd enough to identify the causes and eliminate them at their initial stage itself. External causes are those that are external to the organisation over which the management of the organisation has little control. Government's plans and actions, failure of monsoon which affects agriculture and allied industries, emergence of strong competitors etc., are some of the external factors. Though sickness may be caused either by internal or external factors, sometimes, the management may be able to revamp, its organisation, plan suitable strategies and take on the external factors to reduce their impact.

White Elephants

- Quarterly reports of the Ministry of Project Implementation, Government of India, carry alarming information on project overruns, which have become a serious economic problem. Sixty per cent of the ongoing 132 mega and major projects, individually costing more than Rs 100 crores, and 70 per cent of the ongoing 199 medium projects, costing more than Rs 20 crores each, in the public sector are running behind schedule, with time overruns ranging from 5 months to 13 years. Cost overruns take away about 20 per cent of the total financial outlay on industrial and infrastructural development projects. Of the 331 central projects, 217 have cost overruns of 80 per cent of the original cost.

- A sectorwise analysis shows that die cost overrun is 63 per cent in fertilizer projects, 55 per cent in power projects, 53 per cent in railways, and 48 per cent in chemicals and petrochemicals. In iron and steel projects, of the 11 per cent under implementations, 10 per cent are on cost overrun, while 7 per cent arc suffering from, both, cost and time overruns.
- Such massive overruns not only erode financial outlays on development projects, but also thwart progress in industrial and infrastructural development.
- In India, almost every project in infrastructural sectors has been delayed for one reason or the other.

8.4.1 What is Credit Rating?

Credit rating as 'an evaluation of the financial trustworthiness of a company or individual particularly with regard to meeting some obligation.

There are three credit rating agencies in India. They are:

- Credit Rating and Information Services of India Ltd., (CRISIL)
- Investment Information and Credit Rating Agency of India Ltd., (ICRA)
- Credit Analysis and Research Ltd., (CARE)

The credit rating agencies in India carry out business analysis, financial analysis, fundamental analysis and environmental analysis before assigning rating to a particular debt instrument.

8.4.2 What is Enterprise Resource Planning (ERP)?

The term Enterprise represents and undertaking in which a group of people work with a common goal. The term Resource represents factors like men, material, money, time and the things that are required to run an enterprise and that are consumed during the process of running an enterprise.

The ERP system is a computer software program that supports the ERP.

ERP software is an integrated software program that runs of a single data base in such a way that the various departments can share information easily among themselves and can communicate with each other.

Schematic representation of an ERP system.

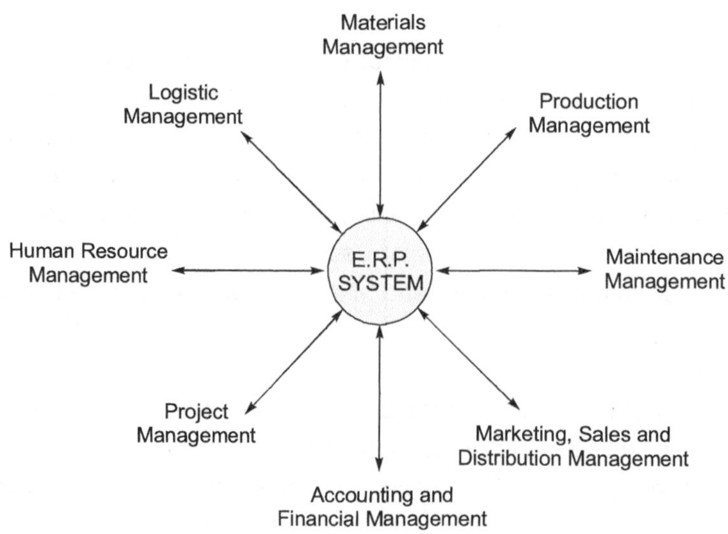

Fig. *ERP System*

8.4.3 What is ABC Analysis?

ABC analysis is based on the principle of 'vital few and trivial many'. ABC analysis segregates the material items that are consumed in a production process/project into three groups, viz., A-items, B-items and C-items.

A-items: We could observe that about 70% to 75% of the total money spent is spent on a few items of materials that account for only about 5% to 10% of the total items of material consumed. Such materials that form part of only 5% to 10% of the total items of material consumed, but that account for a major chunk of the money spent are called A-items.

Cement and steel are the A-items and inventory of these items is to be closely monitored for controlling the cost of the project.

B-items: These are those items that normally form part of about 10% to 15% of the total number of items and account for about 10% to 15% of the total expenditure on the materials. The B-items do not require a rigid control as required for A-items. In the construction of an industrial shed, for example, steels rods and Asbestos sheets etc., will be B-items.

C-items: These are those items that normally form about 70% to 80% of the total number of items but account for only about 5% to 10% of the total money spent. In the construction of an industrial shed, for example, sand, bricks, paint, wood, glass sheets exhaust fans, binding wires, etc., will be C-items.

8.4.4 What is Technical Consultancy Organisations (TCOs)?

These organisations offer technical consultancy for industrial projects. These organisations were established by the All India Financial Institutions in collaboration with the state level financial / development institutions and commercial banks.

TCOs undertake pre-investment studies on the scope for investment in specific industrial sectors. They also undertake pre-investment studies of individual projects in order to assist entrepreneurs.

Some of the activities of TCOs may be summed up as below:

- Conducting industrial potential surveys.
- Preparation of project profiles on promising industrial projects.
- Conducting entrepreneurship development programmes.
- Carrying out market research / survey for specific products.
- Offering merchant banking services.
- Offering consultancy for export-oriented projects.

8.5 VALUE ANALYSIS AND VALUE ENGINEERING

The Society of American Value Engineering (SAVE) defines Value Analysis as under:

Value Analysis is the systematic application of recognized techniques which identify the function of a product or service, establish a monetary value for that function and provide the necessary function at the lowest overall cost.

OR

Value engineering is defined as the systematic, analytical approach of identifying unnecessary costs in construction. Unnecessary costs are those which do not provide any enhanced quality, appearance, utility or improved life. It is estimated that by adopting value engineering techniques, savings up to 5 to 10% on large construction projects can be achieved.

For understanding Value Analysis and Value Engineering, we shall first define the term 'Value'.

$$\text{Value} = \text{Performance} \div \text{Cost}$$
$$= \text{What does a function do?} \div \text{How much does it cost?}$$

- Value Analysis and Value Engineering are the two terms that are used interchangeably. Though both Value Analysis and Value Engineering attempt to identify unnecessary costs with a view to eliminating them, there exists a subtle difference between the two. The difference between the two lies in the time and the stage at which the techniques are applied.
- **Value Analysis is applied for an existing project, whereas Value Engineering applies the same rules to a new project.** Value Analysis is carried out with the objective of improving the value of an existing product. It is a remedial process whereby the unnecessary costs already incurred in a product are identified and reduced. Value Engineering is applied at the design stage with the objective of ensuring that unwanted features are not included in the product. Value Engineering is a preventive process. Value Engineering, when carried out at the initial stage of a product can reduce the cost of production substantially.

Uses of Value Analysis

- Value Analysis reduces the cost of a product.
- It helps employees to have a better understanding of the jobs that they do in the organisation
- Employees become more cost-conscious and value-oriented.
- It creates a conducive climate for the emergence of innovative ideas.
- Implementing changes become easier, since employees become aware of the benefits of Value Analysis and do not offer any resistance to changes that are carried out for the benefit of the organisation.

The following situations in an organisation indicate the immediate need for carrying out Value Analysis:

- There is continuous decline in the sales of the Company's products.
- Rate of return on investment shows a falling trend.
- Competitors offer their products at a relatively lower price to the customers.
- The company defaults in meeting delivery dates.
- Competitors come up with new and improved designs for the products.
- Processing and tooling costs go up steeply.

How to Measure Value?

Value can be measured using Value Index. Value Index (VI) is the ratio of Function Worth to Function Cost.

$$VI = \frac{FW}{FC} \leq 1 = \frac{\text{Aggregate value in given year}}{\text{Aggregate value in base year}}$$

where, FW = Function Worth, FC = Function Cost

VI can take a maximum value of 1.00. When the value of VI is less, it indicates that there is scope for value improvement.

Value Gap

The difference between cost and worth indicates the scope for possible value improvement. When the worth is very much below the cost, it indicates that there is greater scope for value addition. The difference between Function Cost and Function Worth is known as Value Gap (V.G.).

$$\text{V.G.} = \text{Function Cost} - \text{Function Worth}$$

A higher value of Value Gap indicates a greater scope for value improvement.

Previous ESE Prelims Questions

Q.1 ABC analysis in materials management is a method of classifying the inventories based on the
(a) Economic order quantity
(b) Value of annual usage of the items
(c) Volume of material consumption
(d) Quantity of material used

[ESE-2018]

Ans. (b)

Q.2 It is expected to receive Rs. 5,000 annually for 3 years with each receipt occurring at the end of the year. With a discount rate of 10%, the present value of the annuity will be nearly
(a) Rs. 12,435
(b) Rs. 9,945
(c) Rs. 4,975
(d) Rs. 2,487

[ESE-2019]

Ans. (a)

Q.3 A machine is expected to generate cash saving (after-tax) of Rs. 50,000 per annum over a period of 5 years. Salvage value of machine is 40% of the original cost. If accounting rate of return is 20%, cost of two such machines will be
(a) Rs. 78,125
(b) Rs. 1,56,250
(c) Rs. 3,12,500
(d) Rs. 6,25,000

[ESE-2019]

Ans. (c)

Objective Brain Teasers

Q.1 The principal function of a warranty is to:
(a) Provide assurance of the level of quality to be provided
(b) Provide a way to assert claims for late payment
(c) Provide a way to allow additional time following acceptance to correct deficiencies, without additional costs
(d) Ensure that goods purchased fit the purposes for which they are to be used

Q.2 A contract is an agreement between
(a) Two parties valid in law
(b) Several agencies
(c) Three agencies
(d) Two parties without legal binding

Q.3 Lump sum contract is
(a) Commonly followed by many agencies
(b) Followed only by Govt. Departments
(c) Followed only in private contracts
(d) Followed only by local bodies

Q.4 Unit price contract means
(a) Item wise contract
(b) K-2 form of contract
(c) Same as lump sum contract
(d) Both (a) and (b)

Q.5 A turn-key contract means
(a) To design and build
(b) To build the structure only
(c) To perform all the functions from inception to completion of the construction
(d) The same as the unit price contract

Q.6 E.M.D. is
(a) Depositing 2% of the contract value
(b) Same as security deposit
(c) Depositing 10% of the contract value
(d) Depositing only 5% of the contract value

Q.7 The security deposit is
(a) Refunded in the middle of the contract
(b) Refunded after the maintenance period

(c) Not refunded
(d) Refunded as soon as the construction is over

Q.8 An arbitrator is one who mediates between
(a) Two aggrieved parties
(b) Public and workers
(c) Engineer and architect
(d) Public and the contractor

Q.9 A project is engaged in making electronic devices. It is necessary to purchase some materials to make the printed circuit boards. All of the parts are commonly available from several vendors. The most likely contract that should be issued for these parts is:
(a) Unit price contract
(b) Firm fixed price contract
(c) Cost reimbursable contract
(d) Award fee contract

Q.10 With a clear contract statement of work, a seller completes work as specified, but the buyer is not pleased with the results. The contract is considered to be:
(a) null and void (b) incomplete
(c) complete (d) waived

Q.11 Which type of contract is enforceable by law:
1. Oral 2. Written
(a) 1 only (b) 2 only
(c) Both 1 and 2 (d) None of these

Q.12 With regard to the contract, which are correct
1. It is legally binding
2. It is always a written document
3. A contract is legal irrespective of time and place.
(a) All of the above (b) 1 and 2
(c) 2 and 3 (d) 1 only

Q.13 What are different components of contract?
1. Capacity 2. Consideration
3. Offer 4. Legal
5. Acceptance
(a) All of the above (b) 1, 2, 4 and 5
(c) 3, 4 and 5 (d) 1, 2, 3 and 4

Q.14 In a project the need of contract may arise due to
1. Availability of technical know how
2. Non-availability of man-power
3. Lack of budget
4. Avoidance of manufacturing related risk
(a) All of the above (b) 1, 3 and 4
(c) 2, 3 and 4 (d) 2 and 4

Q.15 In ABC analysis most money is spent on which type of items?
(a) *A* (b) *B*
(c) *C* (d) same for all

Q.16 When in a project the order of estimate of cost can be done within accuracy of ±10% of the cost?
(a) at conceptualization
(b) at initiation phase
(c) after the baselining is done
(d) when project is near completion of execution phase

Q.17 What can be reasons of cost overrun?
(i) Time delay (ii) Scope change
(iii) Wrong planning (iv) Inflation in market
(a) All of the above (b) (i), (ii) & (iii)
(c) (iii) only (d) (ii) & (iv) only

Q.18 Which of the following is based on 80-20 principle?
(a) Ishikawa diagram (b) Pareto Analysis
(c) Scatter diagram (d) Monte Carlo analysis

Q.19 A positive cost variance and negative schedule variance means
(a) Project is over budget and ahead of schedule
(b) Project is under budget and behind of schedule
(c) Project is over budget & behind schedule
(d) Project is under budget and ahead of schedule

Q.20 Which type of contracts among following would have greater risk for vendor?
(a) Fixed price
(b) Cost plus
(c) Cost reimbursable
(d) (b) & (c)

Q.21 Which type of contract among following offers greater risk for buyer?
(a) Fixed price (b) Cost plus
(c) Cost reimbursable (d) (b) & (c)

Q.22 **Assertion (A):** Earnest money deposit is forfeited for all unsuccessful bidders.
Reason (R): Earnest money deposit is only 1-3% of the total value
(a) Both Assertion (A) and Reason (R) are individually true and Reason (R) is the correct explanation of Assertion (A).

(b) Both Assertion (A) and Reason (R) are individually true but Reason (R) is NOT the correct explanation of Assertion (A).
(c) Assertion (A) is true but Reason (R) is false.
(d) Assertion (A) is false but Reason (R) is true.

Q.23 What doctrine causes a party to relinquish rights under a contract because it knowingly fails to execute those rights?
(a) Assignment of claims
(b) Material breach
(c) Waiver
(d) Warranties

Q.24 Which term describes those costs in a contract that are associated with two or more projects but are not traceable to either of them individually?
(a) Variable (b) Direct
(c) Indirect (d) Semivariable

Q.25 Of the following which one is NOT true concerning a contract?
(a) It is a legal relationship subject to remedy in the courts.
(b) It can take the form of a complex document or a simple purchase order
(c) It is a mutually binding legal relationship that obligates the seller to provide specific products, services, or results and obligates the buyer to pay the seller.
(d) It includes a specific contract management plan.

Answers

1. (a)	2. (a)	3. (a)	4. (d)	5. (c)
6. (a)	7. (b)	8. (a)	9. (b)	10. (c)
11. (c)	12. (d)	13. (a)	14. (d)	15. (a)
16. (d)	17. (a)	18. (b)	19. (b)	20. (a)
21. (d)	22. (d)	23. (c)	24. (c)	25. (d)

Explanations

1. (a)
A warranty is one party's assurance to the other that goods will meet certain standards of quality, including condition, reliability, description, function, or performance. This assurance may be express or implied.

9. (b)
A fixed price contract will be best in the case of products and to protect buyer risks.

10. (c)
The work is done as specified so it meets the acceptance criteria. Hence it is complete.

11. (c)
Both are enforceable by law if evidence exist.

12. (d)
Contract can be oral.
Control is valid within the boundary where it is signed. As a rule in India right not be same in Europe.

13. (a)
Different components of contracture

C – capacity	C – Consideration
O– Officer	L – Legal
A – Acceptance	

14. (d)
Technical know how helps in making decision to manufacture not to give contract.

16. (d)
It can be seen in the cost funnel that accuracy becomes more precise at later stages of project.

19. (b)
CV → Positive → under budget
SV → Negative → behind schedule

20. (a)
Fixed price contract has more risk for vendors

21. (d)
Cost plus and cost reimbursable would offer greater risk for buyer.

23. (c)
Under the doctrine of waiver, a party can relinquish rights that it otherwise has under the contract. If the seller offers incomplete, defective, or late performance, and the buyuer's project manger knowingly accepts that performance, the buyer has waived its right to strict performance in some circumstances, the party at fault may remain liable for provable damages, but the waiver will prevent the buyer from claiming a material breach and, thus, from terminating the contract.

25. (d)
A contract management plan is not part of a contract. It is used to identify how the contract will be administered.

■■■■